The Dynamics of Innovation

Strategic and
Managerial Implications

Springer
Berlin
Heidelberg
New York
Barcelona
Hong Kong
London
Milan
Paris
Singapore
Tokyo

Klaus Brockhoff · Alok K. Chakrabarti
Jürgen Hauschildt (Eds.)

The Dynamics of Innovation

Strategic and
Managerial Implications

With 33 Figures
and 46 Tables

 Springer

Professor Dr. Klaus Brockhoff
Professor Dr. Dr. h. c. Jürgen Hauschildt

University of Kiel
Institute for Research in Innovation Management
Westring 425
D-24098 Kiel, Germany

Professor Alok K. Chakrabarti PhD
New Jersey Institute of Technology
School of Management
University Heights
Newark, NJ 07102-1982, USA

ISBN 3-540-65659-6 Springer-Verlag Berlin Heidelberg New York

Cataloging-in-Publication Data applied for
Die Deutsche Bibliothek - CIP-Einheitsaufnahme
The **dynamics of innovation** : strategic and managerial implications /
Klaus Brockhoff ... (ed.) -
Berlin; Heidelberg; New York; Barcelona; Hongkong; London; Mailand; Paris;
Singapur; Tokio :
Springer, 1999
ISBN 3-540-65659-6

Hardcover design: Erich Kirchner, Heidelberg
SPIN 10717162   42/2202-5 4 3 2 1 0 - Printed on acid-free paper

Foreword

Jörg Bensinger, a group head of Audi corporation's R&D department, had been waiting for long to find a chance to advertise his idea
of a four-wheel drive for passenger cars to one of the board members. Favorable experiences had been collected in drive tests with the
Iltis, a jeep-like car developed for use in the German army. The experiences showed extremely good performance on icy roads and in
snow. Bensinger's chance came in February of 1977, when he could
talk to Ferdinand Piëch, then R&D vice president of Audi and a
technology buff. At this time Audi wasn't quite considered as a technological leader in the public. Technology based innovations were
expected from Mercedes or Porsche by many customers. Piëch, Bensinger, and others sensed that introducing the four-wheel drive to
passenger cars could initiate a strategic change. Under great secrecy
development work and prototype construction were commissioned.
One obstacle seemed to be space requirements for the gear-box.
Hans Nedvidek, former race-track engineer in the Mercedes team,
was assigned to the team, and he developed an ingenious solution to
the problem. It took until September of 1977 until other board members were informed, and after some rallying the board found a consensus in the next month to authorize further development steps for a
four-wheel drive car.

However, Audi is a subsidiary of Volkswagen Corp. The accord
of the much bigger mother had to be secured. It was thought that this
could best be done by demonstrating what the fourwheel drive could
mean to a passenger car operated under conditions of ice and snow.
Volkswagen board members in charge of marketing and sales were
invited in winter to the steepest Alpine pass road in Austria, which
was mastered by the prototype car on normal summer tires. This lent
support to the idea, and in May of the next year Volkswagen's vice
president for R&D ok'ed the project as well. Another two years later,
in March of 1980, the Audi Quattro Coupé was first presented on the
Geneva Automobile Show.

It had taken three years to push an innovative idea through the organization, to keep it alive, to find the project specialists, and to initiate the necessary strategic change. This helped to position Audi
quite differently than before.

This short case illustrates at least three things. First, innovations might lead to a radical restructuring of a firm. Second, the necessary restructuring is only possible if highly dedicated and committed individuals cooperate to achieve the desired outcomes. They need to stick out their necks and fight opposition, as not every innovation is quite welcome. Third, any successful innovation has to fit into a strategic framework. This, in itself, is difficult to identify and it is in flux, depending on changes in the technological and competitive environment as well as the mastering of the innovation process itself. From the many complex questions that are tied to these processes, we have chosen to investigate in more detail dynamic aspects of strategy formation and strategic impacts on the one hand, and dynamic aspects of project management on the other hand. Results of these investigations were discussed with researchers who have similar interests during a conference held in Hamburg, Germany, in 1998. This volume unites the majority of the papers presented.

Starting with the environment for strategic planning, *Bierly and Chakrabarti* observe its increased volatility, which might lead to the point where firms find themselves in industries without clear boundaries. The combination or fusing of traditional industries is one of these cases, where firms face new competitors, a changing industry structure, new technological demands. This has consequences for the determination of the success factors as well as the ability to integrate different knowledge areas. Advice is given as how to manage a company through industry fusion, focusing on knowledge management and the choice of partners. The pharmaceutical and biotechnology areas are used as examples to illustrate specific ways of managing through industry fusion.

Brockhoff argues that diffusion of technological knowledge variable and the complexity of knowledge going into products span what is called a knowledge spectrum. Firms might take particular positions in this spectrum. However, none of these positions appears to be stabile or undisputed, contrary to some view found in present-day management literature. Therefore, careful analysis of the technological conditions and their possible changes in the knowledge spectrum is necessary to evaluate conditionally preferred positions. No specific path through the spectrum evolves as the most successful strategy.

Organizations develop innovations to adopt to the external environment, or to pre-empt a change in their environment, in order to sustain or increase their effectiveness or competitiveness. Innovation researchers differentiate various types of innovations, such as administrative and technical, product or process, radical or incremental innovations. It can be hypothesized that organizational structure and firm strategy interact with the success of adopting various types of innovations. *Damanpour and Gopalakrishnan* examine the patterns of adoption of the three types of innovations just mentioned. Using theoretical models of innovation adoption they develop a basis for empirical investigations of innovation type adoption. They argue that congruency in patterns of adoption of innovation types facilitates organizational adaptation and enhances organizational effectiveness.

In the next paper, *Leker* tackles a very similar problem. He is interested in measuring performance changes of a firm following strategic changes. It is only in passing that one should note that the concept of strategic change in itself is not easily defined and measured. The present longitudinal study covers five years. During this period it was possible to identify changes of corporate strategy in 65 major German corporations. Using confidential data collected by a bank in its credit ratings, it was possible to study the impact of strategic changes on perceived solvency and on published performance figures. Interestingly, various clusters of firms can be identified that show significantly different reaction patterns.

Yet another approach to illustrate strategic changes is chosen by *Ernst*. Using patent data, he constructs patent portfolios for a number of competing firms and the technologies that they employ. Developing a series of patent portfolios for subsequent years, it is possible to trace the changes of a firms' patent positions. This can be helpful both to identify and to guide R&D strategies or to identify technological core competencies.

The final paper in the first part serves as a bridge to the second part. Research done in Kiel by Stephan Schrader, who unfortunately died at a very young age in 1997, identified relationships between strategic choices of resource use, company performance, and educational background of U.S. CEOs. To have a comparative study done for German board members was considered important, since institutional differences in companies and differences in educational systems, among others, prohibit an immediate transfer of the findings to Germany. Such a study was performed by *Harhoff*, a friend and former colleague of Schrader. He shows that educational background impacts on resource use, with engineering people being employed in companies with more continuous and more intense R&D. Furthermore, managers with a business administration background work in companies with higher growth rates. It remains to be studied what is cause and what is effect. The results alone show the strategic importance of education, strategic choice, and company performance.

In the second part of these proceedings the contributions deal with the management of innovation projects. *Hauschildt* gives an overview over 25 years of research on championship in innovation. Champions or promotors are individuals who support innovations enthusiastically. Research initially proved that an actively committed champion is a most important factor for success in the management of innovations. At the same time, however, a variety of other individuals were observed who were also striving to make the innovation successful. There are now a number of explanations which support the notion that the successful effect of these individuals is due to their skills in dealing with conflicts constructively and handling information creatively. A troika structure securing the roles of power promotor, process promotor and technology promotor in particular is the most successful structure for the internal management of typical innovation projects. It is possible that the increase in inter-firm innovation activity will shift the role of the process promotor more towards that of a relations promotor. Dynamics of role change are important research issues.

Markham, in his contribution, shows empirically how champions influence others to support their projects. His investigation reveals that standard influence tactics are not successful at generating support. Champions are successful at leveraging personal relationships to gain target compliance and willingness. Yet, these data did not reveal a champion's positive contribution to project performance from the point of view of the team. Champions, however, maintain a more optimistic view of the project than do team members.

Hauschildt, in his second contribution, presents the findings of an empirical survey on opposition to innovation. Innovations are not always welcomed by those affected by or involved in them. They often trigger resistance. This study shows types, intensity, and effects of such opposition. The basis of the study is a questionnaire survey of applicants for the 1998 German Industry Award for Innovation. The sample contains 154 projects by 151 companies. Two variants of opposition are identified by factor analysis: destructive and constructive opposition. Constructive opposition aims at changing the innovation, destructive opposition aims at preventing or delaying it. Promotors counter destructive opposition effectively.

Gemünden et al. analyze the impact of the start-up conditions of European multi-partner research projects, such as ESPRIT-projects, on the development and the success of the projects. The conceptual model proposes partner-fit comprising social fit (trust and commitment), resource fit (task-competence and complementarity), and goal fit (goal clarity and goal consensus), as major determinants of project success. The success measures include overall success, effectiveness, efficiency, collaboration, and learning success. The basic hypothesis is: The better the fit of the partners at the beginning of the project, the better will be the development and the results of the project. The data analysis confirms their hypothesis.

Kessler and Chakrabarti state that there is disagreement on how concurrent development affects innovation speed, new product quality, and development costs. To clarify these relationships, they examined new product development projects in a variety of industries. Regression results reveal that (a) concurrent development significantly influences the three project success indicators, and (b) it has mixed effects on these indicators. A higher degree of concurrentness is found to increase innovation speed, but it also tended to increase development costs and decrease product quality. This suggests that firms need to adopt more balanced or optimization related views of concurrent development and they should avoid "shotgun" approaches to speeding up development. It may also imply that there are better and worse ways to execute concurrent development. In other words, managers need to consider potential trade-offs in matching R&D strategy with their development objectives.

Finally, *Farris* focusses on high-impact innovations. These are innovations that have had a disproportionately great impact on standards of living, job creation and economic growth. Typically, only a very small number of innovators has contributed work of such significance to their companies. Examples include the cellular telephone, geostationary satellite, minicomputer, personal computer, environmentally friendly herbicides, and commercial jet aircraft. How could the success of these people be explained, and could it be managed? To shed light on these questions, interviews were held with people recognized nationally for their critical roles in developing these key technologies, and questionnaires were administered to engineers and scientists whose work was judged to be of greatest value to their organizations. Based on this information, tentative conclusions can be drawn about key success factors.

The need for more research into the questions of innovation dynamics is evident. It was very fortunate to find a group of American and German scholars who concentrated their efforts on empirical research. This led to a grant application to the German-American Academic Council by the editors of this volume. The Council provided some funding through its TransCoop Program that helped to coordinate the research, to discuss interim results and to have the final, two-day conference in Hamburg. This funding was matched by the institutions to which the editors are affiliated, New Jersey Institute of Technology's School of Management and the University of Kiel's Institute for Research in Innovation Management. The conference attracted other researchers who are interested in the study of innovation dynamics, such that a critical debate on this field emerged. This will certainly lead to more research. It is in this sense that we may call the present papers interim reports. They are presented both as a documentation of the work done and as a stimulus for others to join in the fascinating research area which the authors of this volume could explore only to a limited degree. Literature from all papers has been collected at the end of the book. It may well serve as a bibliography that makes European articles and books more readily accessible to Anglo-American readers.

To all the institutions that supported this work, and to all contributors who helped to make this a stimulating research experience go our sincere thanks. Special thanks go to *Ms. Dörte Jensen*. She had the arduous task of producing a camera-ready manuscript from piles of papers in very different formats. We could not have succeeded in editing this book without her help.

Kiel, Germany, and Newark, New Jersey, in December, 1998

Klaus Brockhoff, Alok K. Chakrabarti, Jürgen Hauschildt

Contents

Part I: Strategic dynamics

Managing through industry fusion

Paul Bierly / Alok K. Chakrabarti

Managing through industry fusion

Paul Bierly / Alok K. Chakrabarti

1. Introduction

Many industries are becoming "boundaryless" and associated with rapidly changing competitive environments (Bettis, Hitt 1995; Hamel, Prahalad 1994; D'Aveni 1994). The increasing rate of technological diffusion and the globalization of markets are two of the main reasons why firms' environments are becoming more dynamic. New technologies are diffusing across industry boundaries at an increasing rate primarily because of the tremendous increase in access to information through the advances in computers and communications. This "sharing" of technologies among industries creates new competitors from surprising sources. Additionally, the globalization of almost all major industries has increased the rivalry, and hence the rate of innovation, in many industries by introducing players with different knowledge sets, different perspectives of the future, and a different set of industry norms. Partners can be very valuable in this type of setting to help better understand the changing rules of the game (Ohmae 1989).

These changing environmental conditions can create a condition where firms find themselves in industries without clear boundaries as their traditional industry is combined with one or more different industries. This combination, or fusing, of different industries is a condition that is referred to as industry fusion (Bierly, Chakrabarti 1998). A firm whose industry is being fused with another is in a difficult position because it is facing new competitors and a changing industry structure. This appears to be a new phenomenon created by the changing competitive landscape and many industries have not yet entered the industry fusion stage. However, it appears that this will be an increasingly common occurrence in the near future. Examples of industries currently in the industry fusion stage are

(a) the integration of the computer, communications, and enter-
    tainment areas,

(b) the integration of the pharmaceutical industry (skill in organic
    chemistry) with both the biotechnology industry (skill in mo-
    lecular biology) and the pharmaceutical distribution industry,
    and

(c) the integration of the many different types of financial services.

In each of these cases there are several critical knowledge areas
that are being combined together in complex ways such that the future industry structure will be completely different than the current
state. Furthermore, the products of the future are envisioned to be
much different than they are today and will rely on knowledge areas
that are only just starting to be developed.

Under conditions of industry fusion, traditional approaches to
strategic management appear to be inadequate. Specifically, the
school of strategy that Mintzberg (1990) refers to as the "positioning" school, which was pioneered by Porter (1980), relies on techniques that are inappropriate during industry fusion. Conducting an
industry analysis using the Porter Five Forces model, analyzing
one's direct competitors, or establishing and defending a market
position can be myopic activities that fail to account for the dramatic
changes that lurk ahead. These traditional approaches fail to consider how technological advances, particularly radical and non-linear
advances, quickly change industry structures and make "positioning"
competitive advantages vulnerable (Lei et al. 1996; D'Aveni 1994).
Industry fusion intensifies an already chaotic condition of technological change by introducing new, very powerful competitors from
other industries and the history of learning that has evolved in this
other arena. The competitors from different industries follow different industry recipes for success (Spender 1989) providing top managers with a different perspective of competition and a different
sense of what is considered "fair play". Predicting how technologies
will evolve during industry fusion is particularly challenging because it is often the case that many complementary technologies
must evolve together. Frequently, these complementary technologies
are developed by very different types of firms and it is often difficult
to understand and integrate the different technological areas.

The goals of this paper are as follows. First, we intend to illustrate
how industry fusion changes the dynamics of competition. There is
the introduction of new players with different perspectives who fol-

low a different set of rules; there are new critical success factors mostly based on the ability to develop and integrate different knowledge areas; and competition becomes more complex and takes place on several platforms. Second, we will offer general guidelines how to effectively manage a company through industry fusion, focusing on the importance of knowledge management and the use of partners. Third, we will use the pharmaceutical/biotechnology arena as an example to illustrate specific challenges of managing through industry fusion.

2. Industry fusion and the dynamics of competition

2.1 New players

During industry fusion, the most apparent change to a firm's competitive environment is the introduction of many new, potentially very powerful competitors. This dramatic increase in the number of new players is much different than the gradual flow of new entrants into most existing industries. In most established industries, entrants are usually smaller than the industry leaders, have less bargaining power with suppliers and buyers, and often follow a niche strategy to avoid direct confrontation with the leaders. Under such conditions, the established players in mature industries have many inherent competitive advantages over new entrants including brand recognition, access to distribution channels, economies of scale, and economies of learning (Lieberman, Montgomery 1988). In most mature industries, these advantages are insurmountable and the established market leaders maintain dominant positions for a long period of time (Grant 1998). This is particularly true in an industry with a strong dominant design and high switching costs (Teece 1987).

During industry fusion, firms encounter new players that are much different than their competitors of the past. Some of these new competitors were the dominant players of a different industry, with expertise in different knowledge areas. They probably pursued different avenues of research and development, developed expertise of different supplier and distribution networks, learned different manufacturing techniques, and knew how to compete in a different competitive landscape (i.e., faster or slower rate of technological change, more or less government regulation). Thus, firms may find themselves at a competitive disadvantage relative to their new competitors in several portions of the value chain.

Spender (1989) illustrated that within an industry an industry recipe develops that can be viewed as a shared set of ideas that become institutionalized and provide a guide to action. The industry recipe can also provide norms for competition so that rivalry does not become destructive and industry profits remain high. For example, different signaling techniques may become commonplace for changes in capacity, price, etc. During industry fusion, the competitors that previously were in different industries will be following different industry recipes and will be following different rules to the game. Thus, competitive actions and responses will be much more difficult to interpret, creating a more chaotic condition.

2.2 New critical success factors

What it takes to be successful during industry fusion may require quite different capabilities than what it takes to be successful in an established industry. First and foremost, to successfully manage through industry fusion, firms must have a clear vision. The underlying assumption for all of its strategic actions will be how top managers envision the future products, customers and competitors after industry fusion. Part of the clarity of a firm's vision will be dependent on its technological know-how and its ability to forecast how key technologies will evolve. However, much of the vision will also rely on instincts and intuitive feelings. For example, Motorola's vision of the future of the telecommunication/computer/entertainment arena is an interrelated complex system where all data is transferred via a sophisticated satellite system. Almost all of the investments in capital and the development of core competencies follow from this basic belief. However, many of its competitors envision most data being transferred by fiber optic cable and believe a satellite system will only be a secondary media. Having a clear vision allows a firm to most efficiently leverage its resources and capabilities, but the accuracy of the vision is what will determine the effectiveness of its long-term strategy. Motorola's success will depend partly on how well it (along with its partners) can develop its satellite system, but also on how well competitors develop alternative systems.

A second critical success factor for firms going through industry fusion is their ability to integrate different knowledge areas (Bierly,

Chakrabarti 1998). The integration of knowledge areas requires efficient communication (encoding, transmission, and decoding) between the experts of the specific areas. The more tacit the knowledge area, the more difficult it will be to communicate and integrate with other areas. In general, knowledge areas may refer to the skill and know-how of many different subjects, including technological and administrative knowledge, expertise of different functional groups (and different specialties within functional areas), knowledge of product and process design, and knowledge of market conditions. Whether a specific knowledge area is either developed internally or acquired externally, a critical issue is to determine what level of understanding of a knowledge area is required by either non-specialists or specialists in another area if the knowledge area is to be successfully integrated into a specialized knowledge area. In other words, how well does the firm (or sect of the firm) have to understand an external knowledge area if it is able to use it? Cohen and Levinthal (1990) claim that for most external knowledge areas to be used effectively, the organization must already possess a considerable level of competence in the area so the value of the technology can be recognized, assimilated, and applied to commercial ends. They refer to this notion as the firm's "absorptive capacity". During industry fusion, knowledge areas from two or more industries must be combined. The firms that can do this best will have a distinct competitive advantage over the other firms. Kogut and Zander (1992) refer to this type of competitive advantage as the "combinative capability" of a firm. This type of capability is particularly challenging during industry fusion because it requires absorptive capacity in very different knowledge areas. In general, firms that do this well will have a broad knowledge base, have an open "learning" environment, and will have experience communicating across disciplines (i.e., cross-functional teams).

A third critical success factor for firms going through industry fusion is their ability to work well with partners and to manage a network of players. It is probably unrealistic to expect a single company to be able to aggressively pursue each of the diverse knowledge areas that are integrated during industry fusion due to the vast amount of resources that would be required to do so. It would also probably be unwise to do so, even if a firm is able, since it could cause confusion and deterioration of core knowledge areas. For example, AT&T attempted to become a leader in many of the critical knowledge areas associated with the fusion of the telecommunications and computer industries throughout the late 1980's and early 1990's. However, management had difficulty maintaining control over such a diverse group of knowledge areas and they started to lose their competitive advantage in several key areas. In 1996, they decided to reverse their strategy, and split the company into three smaller units so that each group could remain focused in its narrower cluster of core competencies.

Thus, firms must rely on strategic alliances and networks to support their knowledge base (Grant 1996; Hamel 1991; Powell et al. 1996). Therefore, a critical success factor for firms going through industry fusion is their capability of successfully managing these alliances. Specifically, this requires skill in communications, conflict resolution and coordination (Mohr, Spekman 1994). Additionally, it is critical that the firm develops a reputation of being trustworthy (Barney, Hansen 1994).

2.3 More complex competition on several platforms

Industry fusion is often characterized by (a) competition in a wide array of knowledge areas, (b) competition between networks and (c) competition within a network. The superior product in the future will integrate leading technologies from each of the critical knowledge areas. A firm or network that is not competitive in any one of the critical technologies may be blocked from competition. Thus, there is direct competition in each of the knowledge areas and in the ability to integrate the different knowledge areas.

Competition among networks is much more difficult to analyze for several reasons. First, the members of a network can change much faster than a single firm could change its own capabilities. Overnight, a network can change a glaring weakness into a strength by the addition of a new team player. Competitive analysis becomes a much more dynamic process. Second, the most sustainable competitive advantage of a network may be very hard to imitate. As discussed earlier, the most critical capability of a network may be its ability to integrate different knowledge areas. How a specific network successfully accomplishes this task is probably difficult to explicitly determine. Much of the communication in the process will be informal and tacit. Third, comparing leaders in each specific knowledge area may be misleading because many of the technologies may be complementary in nature and influence development of related technologies. A superior position in a specific knowledge area may be wasted if the complementary areas that are needed to implement the advances are not simultaneously developed (Teece 1987). For example, a revolutionary new computer software program may not be able to be utilized until the current state of hardware is advanced. (Note, the reverse could also be the case where the hardware is limited by the software.)

Additionally, there is frequently competition within a network. Even though the firms within a network are partners, they still jockey their positions trying to improve their strategic leverage relative to each other in an attempt to maximize their profits. Ideally, a firm usually wants to be the center node of the network, which is usually the most powerful position. Power in the network typically is a function of the firm's size, reputation, and expertise in a critical knowledge area. Network centrality and power is often redistributed after different industry events (Madhavan et al. 1998).

3. Managerial functions

3.1 Managing knowledge base and strategies

It is obvious from our preceding discussions that managing the knowledge base of the organization is an important strategic component of the overall corporate strategy for success. It is important here to make a distinction between the tacit and explicit knowledge (Nonaka 1994). The knowledge strategies must include both types of base knowledge.

Although technology has been mentioned as an important component of corporate strategy, little attention has been paid on the empirical data that could lead to effective technology strategy. Maidique and Patch (1982) identified six dimensions in technology strategy:

(1) level of specialization in technology selection,
(2) level of technological competence,
(3) source of technological capability,
(4) flexibility of R&D policy and structure,
(5) level of R&D investment, and
(6) competitive timing.

In their longitudinal study of six major chemical firms in the US, Sen and Chakrabarti (1986) documented the influence of technology in the corporate strategies and concomitant performance. Brockhoff and Chakrabarti (1988) identified four different technology strategies among the German companies. These strategic clusters are:

A. Defensive Imitator: Firms following this strategy emphasize imitative behavior in their technology development program.

B. Process Developer: Firms following this strategy are internally oriented and emphasize process development as their competitive edge.

C. Aggressive Specialist: R&D specialization in certain technological fields is the key component of the strategy.

D. Aggressive Innovators: Firms following this strategy have a broad technology base and a balanced portfolio of product and process innovations.

Knowledge strategies, as defined by Bierly and Chakrabarti (1996a) go beyond what is meant by technology strategy in Brockhoff and Chakrabarti (1988). Knowledge strategy involves strategic choices to be made in the following dimensions:

1. Trade off between internal vs. external learning

2. Trade off between radical vs. incremental learning

3. Speed of learning

4. Breadth of learning.

These four dimensions of knowledge strategies can be manifested in many business decisions involving mergers and acquisitions, joint ventures, formation of strategic alliances, licensing, patent policies, R&D expenditures, product portfolio, etc. Bierly and Chakrabarti (1996a) identified four strategic groups among the US pharmaceutical companies in ethical drugs. They are:

1. Innovators: These firms have highest level of internal learning through R&D, focus on both radical and incremental learning, and are fastest learners.

2. Exploiters: These firms are effective users of technology available and have a broad, but shallow, knowledge base. They have low R&D investment and focus on incremental learning.

3. Explorers: They are characterized by their proclivity to attempt to "hit the home run" with a new block buster drug. They maintain a balance between external and internal learning, but are less aggressive learners.

4. Loners: Firms in this category are ineffective learners, although they do spend money in R&D. They are too focused in narrow technical areas and not able to integrate different streams of knowledge.

Both the technology strategy (Brockhoff, Chakrabarti 1988) and the knowledge strategy (Bierly, Chakrabarti 1996a) have some significance in managing a firm involved in industry fusion. These strategic concepts should help focus a firm in acquisition and integration of knowledge. The concept of "core competence" as proposed by Hamel and Prahalad (1994) provides a compelling argument to narrow the technical fields on which a firm should focus. However, a firm needs to be flexible and adaptive to advances in different but related fields of technology to avoid "core rigidity" (Leonard-Barton 1995). This becomes particularly an important consideration when the industry boundary conditions disappear.

3.2 Integrating different knowledge areas

Other studies in the pharmaceutical industry have shown that corporate performance is linked with the ability to integrate the different knowledge streams (Henderson, Cockburn 1994; Pisano 1994). According to Henderson and Cockburn (1994), "architectural competence - the skill of integrating a wide range of disciplines- and specific expertise in any of these disciplines provide a source of advantage in drug research productivity". The capacity to integrate different knowledge streams becomes important in both internal and external knowledge sources. Pisano (1994) illustrated the importance of integration at different stages of drug development.

Integration of knowledge across disciplinary and organizational boundaries requires effective interface management. Effective interface management may involve structure and management process involving both personal and impersonal instruments (Brockhoff et al. 1996). The following issues should guide the choice of the management instruments:

1. Level at which the interface problem occurs.
2. The type of exchange that should be considered.
3. The reason for creating an interface has to be considered.
4. Task characteristics, such as complexity, frequency, repetition, standardization, etc. are to considered.

Interface at the project level often involves uncertainties about both the ends and the means. An explicit discussion about the various uncertainties may facilitate interface among the various organizational entities involved. This will specifically help identify the information needs.

Integration of knowledge across different boundaries requires the contribution of special people such as champions, sponsors etc. According to Chakrabarti and Hauschildt (1989), "during the life of a project different people assume different informal and semiformal roles which supplement and complement the communication channels and decision making loci". The effectiveness with which people in these roles perform will determine the success of the innovation.

As Brockhoff et al. (1996) have shown, these roles change over time as the scope of an innovation project is broadened with increasing participation of different organizational units. The role requirements are different for these positions. For example, the champion of an innovation must have high level of tolerance for ambiguity, excellent communication skills, political astuteness and a good understanding of the strategic issues. A technical expert needs good technical knowledge and many not require those other skills to the extent, as the champion needs.

3.3 Broadening the knowledge base using partners

Firms often face the dilemma of broadening the knowledge base as opposed to focus on a few product and market areas. Hamel and Prahalad (1994) have provided a very compelling case for focusing on the core competence of an organization. This becomes particularly important when a firm has to allocate its scarce resources among various projects and entities. During the late seventies and early eighties, we have seen a fast trend in mergers and acquisitions leading to the growth of diversified conglomerates. The idea of core competence has since then helped the managerial trends of downsizing, diversification and outsourcing. Leonard-Barton (1995) has voiced the concern over "core rigidity" which is derived from overly excessive dependence on the "core competence". Her argument is that "core rigidity" would stifle innovation and ability for learning. Her argument appears to be more important under conditions of industry fusion when the boundaries of knowledge base become amorphous.

External learning becomes an important means to best exploit the broader knowledge base. There are many mechanisms to achieve this:

1. Strategic alliances: Many companies, particularly in the biotechnology industry have developed strategic alliances to broaden their knowledge base at a moderate level of risk.
2. Licensing: Technology licensing arrangements have provided ready access to external sources of technology and knowledge.
3. Joint venture: By developing joint venture through equity participation, firms have developed access to knowledge base, both technical and non-technical. Joint venture provides various types of opportunities for technology transfer.
4. Mergers and acquisitions: Mergers and acquisitions are quite popular for expanding one's knowledge base. In telecommunication industry, one observes the trend of mergers and acquisitions.

3.4 Importance of the firm's past

The firm's past strategy is an important determinant in its future success. The past strategies determine the technical trajectory it is most likely to follow. Eastman Kodak has long followed the strategy of using photochemical process as its core technology. Shifting its strategy to electrostatic and electronic imaging has been quite painful with dubious effectiveness. Motorola's dependence on analog technology has been quite costly, as digital technology has become more popular. Past strategy and practices also determine the culture of the organization and its receptivity to learning.

Firms are capable of changing their strategy as we have observed in our study of the US pharmaceutical companies (Bierly, Chakrabarti 1996a). We have also noted that certain trajectory can be beneficial for corporate performance. For example, the firms that changed to become "innovators" showed better performance than others did. It takes a conscious management decision to implement a specific trajectory of strategic change.

4. The pharmaceutical/biotechnology area

4.1 Changing environment - industry fusion

To illustrate the dynamics of industry fusion, we will discuss the fusion of two industries: the pharmaceutical industry and the biotechnology industry. During this industry fusion, the knowledge base of the pharmaceutical industry, which is based on knowledge in organic chemistry, needs to be integrated with the knowledge base of the biotechnology industry, which is based on knowledge in molecular biology. In both of these dynamic industries, industry fusion is also occurring in other areas. The biotechnology industry is also fusing with the chemical, agriculture, waste disposal and energy areas; the pharmaceutical industry is also fusing with the pharmaceutical distribution industry. In each of these cases there are several critical knowledge areas which are being combined together in complex ways such that the future industry structure will be completely different than the current state. Furthermore, the products of the future are envisioned to be much different than they are today and will rely on knowledge areas that are only just starting to be developed.

In the past, pharmaceutical companies, relying on their expertise in chemistry and pharmacology, used random screening to discover an effective new drug. Thousands of different chemical compounds would be tested to see if any would have a desirable effect, with minimal adverse side-effects, combating a specific disease. Usually the focus of researchers was on the symptoms of a disease; often how the drug actually worked was not well understood. Success in this environment was based on knowledge of organic chemistry and pharmacology and the ability to efficiently screen thousands of chemical compounds. Typically, a pharmaceutical company would test over 10,000 compounds to find a single drug that may work. Capabilities in marketing and large-scale manufacturing, access to a large distribution system, and name recognition were also critical success factors. The integration of knowledge across the different activities was not particularly important; it was not necessary to fully understand why a group working in another area made their specific decisions.

Since the early 1990's, almost all drugs have been discovered by following a rational drug design, an approach that relies on the structural analysis of target molecules and the deliberate design of agents that affect their function. Many successful drugs are enzyme inhibitors that block specific receptors associated with a certain disease. The inhibitor can be viewed as the "key" that fits the "lock", which is the receptor. Thus, if one understands the detailed nature of the receptor, than inhibitors can be designed that will block the functioning of the receptor. Essentially, the researcher becomes the locksmith who must find the best key for the lock. The successful locksmith must have a strong background in molecular biology, to understand the design of the receptor, and in chemistry, to understand how different drugs will react with the receptor. Indeed, rational drug design requires the integration of knowledge from many different disciplines, including molecular biologists, biochemists, physiologists, chemists, pharmacologists and experts in very specialized fields related to the specific drug. In other words, successful drug design requires the combination of the traditional knowledge base of pharmaceutical companies with the knowledge base of biotechnology companies.

Powell et al. (1996) characterized biotechnology as a competence-destroying innovation, from the perspective of established pharmaceutical companies. For the established pharmaceutical companies to be successful at the rational drug design process, they must "unlearn" the processes associated with random drug screening. What Tushman and Anderson (1986) described is typical of most competence-destroying innovations, the competitive advantages of the established players are diluted and may actually become liabilities. The door is opened for new, smaller players to enter the field who are not wedded to the "old" technology. In the case of drug development, the door is opened for the hundreds of small biotechnology firms to compete with the previously unapproachable pharmaceutical firms. The biotechnology firms have the advantage of being the leaders in research, especially at the basic science level. However, the biotechnology firms also have the same competitive disadvantages as many other players from emerging areas: they lack capabilities in large-scale manufacturing and marketing. Additionally, even though Tushman and Anderson (1986) illustrated this rarely occurs, the established pharmaceutical companies appear to be dramatically transforming themselves in response to this competence-destroying innovation, primarily due to the strong leadership of technically sophisticated senior management (Zucker, Darby 1997).

4.2 Changing industry structure

Over the last several decades, the pharmaceutical industry was one of the most profitable sectors in the U.S. economy, primarily due to its attractive industry structure (McGahan 1994). The industry structure was stable with large companies like Merck, Lilly and Abbott maintaining leadership roles for many years. Direct competition was mitigated by the arrangement of each company focusing on a few therapeutic classes and the effectiveness of the patent system. Buyers did not pressure companies to lower prices, allowing the companies to maintain high profit margins. Additionally, industry entry barriers were high, limiting new competition. Specifically, the established firms spent very large sums on both R&D and marketing. The Office of Technology Assessment of the U.S. Congress estimated that the average amount spent on R&D for each new drug introduced in the 1980's was over $200 million and the drug approval process usually took about 12 years (Office of Technology Assessment 1993). In the U.S., the drug approval process conducted by the Food and Drug Administration (FDA) included three phases of clinical trials: Phase I tested clinical safety, Phase II tested drug efficacy, and Phase III tested for adverse effects from long-term use. Incredibly, most pharmaceutical companies spent even more money on marketing than R&D, developing strong ties with the doctors and pharmacists. Thus, the "old" industry recipe for success was that bigger was better - spending large amounts in R&D and marketing would lead to high profits. This rationale led to mergers of large pharmaceutical firms (e.g., Bristol-Myers and Squibb, SmithKline and Beecham) with the stated purpose being to gain size in both R&D and marketing.

However, the pharmaceutical industry has undergone dramatic changes recently. Two of the most significant changes to the environment have been (a) increased price pressure by managed care networks and the government, and (b) the change from random drug screening to rational drug design, described above. The increased price pressure has forced firms to focus on improving efficiencies throughout the organization. Effort placed on improving operating efficiency in manufacturing and distribution help to reduce costs, but the primary area where large cost savings can be achieved is the improvement of R&D efficiencies. Some pharmaceutical companies have also attempted to reduce this price pressure by acquiring their own managed care providers, led by the Merck acquisition of Medco Containment Services.

The use of a rational drug design approach to new product development has changed what it takes to be successful in the drug industry. Now, the critical success factors are a broad knowledge base and the ability to integrate the broad knowledge base (Bierly, Chakrabarti 1996a). Specifically, pharmaceutical firms must have access to and the ability to use the tools and knowledge set of the biotechnology firms. All pharmaceutical firms have realized that this integration of the skills from the biotechnology area is required for future success. Some firms, like Eli Lilly, SmithKline Beecham and Merck, have actively pursued biotechnology for many years, but others, such as Abbott, Upjohn and Warner-Lambert, have just recently realized that it is not a feasible strategic position to have no ties to biotechnology. For example, Bristol-Myers Squibb, historically a passive player concerning biotechnology, recently established an External Science and Technology group within its R&D division to provide access to external developments. In a short time period the group formed strategic alliances with four biotechnology companies: Cadus Pharmaceutical Corp., SEQ Ltd., Genzyme Transgenics and Agracetus Inc.

The simultaneous need to make R&D more efficient and the need to have access to a broad array of knowledge areas have increased the reliance on external sources of knowledge. Since these technological fields are evolving at such a fast rate and involve different resources, it is very difficult to remain cutting-edge in the wide array of knowledge areas by developing all of the technologies in-house. It is more efficient and allows for more strategic flexibility to work with partners (Bierly, Chakrabarti 1996b). Acquiring other companies or relying on partners can be effective substitutes to innovation (Hamel 1991; Powell, Brabtley 1992; Barley et al. 1992). A mid-range position between these two options, an equity position in the partner, can be effective in helping transfer knowledge and increasing control, while still allowing the biotechnology company to keep its autonomy (Pisano 1989; Bierly, Kessler 1998). Relying on external learning allows companies to focus more on what they do best, without being shut out of advances in other areas. However, relying on external learning also makes the company more vulnerable. Indeed, at no time in the past did pharmaceutical companies entrust the development of critical technologies with outsiders to this extent.

Some pharmaceutical companies (e.g., Lilly) have attempted to integrate the biotechnology and traditional chemistry knowledge areas by the acquisition of small biotechnology companies and the subsequent development of an in-house biotechnology division. The advantages of this approach are that

(a) the pharmaceutical company maximizes its control over the smaller biotechnology company,

(b) the pharmaceutical company is able to land the top scientists of the biotechnology company (assuming they do not leave), and

(c) this arrangement allows for the smoothest transfer of knowledge, since the biotechnology scientists will be working directly with the researchers of the pharmaceutical company (Pisano, 1991).

However, these attempts have frequently been unsuccessful for a variety of reasons. First, pharmaceutical and biotechnology companies are difficult to integrate because they have dramatically different cultures, organizational structures and control mechanisms. Biotechnology companies tend to be much smaller and entrepreneurial with loose, decentralized structures designed to promote creativity and brainstorming. Pharmaceutical companies are more hierarchical and their research projects are more focused and calculated. These organizational designs are both appropriate because both technologies are different in nature: biotechnology is a more abstract, emerging technology, whereas chemistry is more mature. Second, after a pharmaceutical company buys a biotechnology company, many of the most talented researchers initially with the biotechnology firm leave to avoid working for such a large company with a rule-driven atmosphere. This is a major problem since the value of the biotechnology company is primarily in the intellectual capital of the employees. Third, the acquisitions are expensive and may not produce the desired results because the biotechnology firm may be working on issues that are only applicable to a narrow area of research and can not be expanded to other, more profitable areas. Fourth, since many pharmaceutical companies do not have expertise in the biotechnology area and the knowledge associated with biotechnology is mostly tacit and abstract, they overestimate the likelihood of new drug development.

For these reasons, almost all pharmaceutical companies rely on a complex network of partners to develop these diverse knowledge areas, even if they have their own large internal R&D facility (Barley et al. 1992; Powell. Brantley 1992). Expertise in specific knowledge areas can be accessed through biotechnology companies, university researchers, and consultants. Nowadays, there are even several deals with two large pharmaceutical companies doing joint research (e.g., Upjohn and Tanabe), which was rarely done before. Thus, the industry recipe for success has changed. Bigger is no longer necessarily better. Success goes to the company that can (a) access expertise in key, critical areas, (b) understand the significance of major breakthroughs in each of these areas, and (c) has the capability to integrate these different knowledge areas. The integration stage is particularly challenging because the experts in the different fields frequently speak a different technical "language", have different educational backgrounds, and interact infrequently. Since there is a limited number of firms with access to some specific technologies, there is competition to form alliances. Reputation and a perception of trustworthiness are important attributes to attract partners. Firms that act opportunistically and are not trustworthy may find it difficult to find partners in some critical knowledge areas, which may be detrimental to effective innovation (Powell 1998). Randall Tobias, Chairman and CEO of Eli Lilly and Company, summarized the beliefs of many in the industry concerning the importance of partnerships in his 1998 Keynote Address at his company's Annual Shareholder's Meeting:

"It is Eli Lilly's intent to maintain our top-tier position and to keep raising it to the next level of performance. We have pledged to do this by outgrowing our competitors through a constant stream of innovation. What that means, very simply, is that we've got to generate new molecules, and launch attractive new products, at an unprecedented pace ... Incremental improvements alone will not be enough to sustain top-tier performance. Real innovation is what counts ... We will continue to grow this in-house capability, but naturally, we must invest selectively. The life sciences today are so rich with potential that no singe entity can hope, or afford, to own all the new tools, the techniques, the ideas that are available. That is one of the reasons I continue to believe that size alone does not impart a competitive advantage in this industry ... The path we have chosen aims to enhance and complement our own skills, to take Lilly's proven ability to innovate and to multiply it dramatically. That is why we are pursuing strategic collaborations with other scientific organizations. Lilly has one of the richest "partner portfolios" in the industry. To date, we are engaged in more than 100 research and licensing agreements, 37 of which were initiated in 1997 alone. But this move toward increased collaboration is a new and important strategy in our drive for constant innovation. Within just a few years, we anticipate up to half of our new molecules will originate beyond our own walls. This projection, far from implying weakness in our scientific capabilities, actually serves to validate Lilly's expertise. Our world-class scientists are a magnet for other researchers with great ideas ... Once we've established our partnerships, we work to make them better. True collaboration requires a degree of openness and risk-taking that, in the past, might have made us uncomfortable. But we're learning to be faster, smarter and less bureaucratic, and this has been noticed not only by our partners but by our own employees as well. The ability to be a good collaborator, a partner of choice, if you will, is rapidly becoming a key competitive advantage for Lilly ... Collaboration and partnering add greatly to the richness of Lilly's scientific base... they are subtly reshaping our culture, making us a company that's more flexible, more open and more willing to take intelligent risks."

4.3 Strategic response: The case of Ciba Geigy and Chiron

The case of Ciba-Geigy and Chiron illustrates how one company deals with difficulties associated with industry fusion. Ciba-Geigy (which recently merged with Sandoz to become Novartis) was a large Swiss pharmaceutical company with a conservative, formal culture. Since the early 1980's Ciba-Geigy identified the need to develop expertise in biotechnology to complement its strength in the traditional pharmaceutical fields of research. However, they were continually unsuccessful in attempts to develop working relationships with small US biotechnology companies. Specifically, they repeatedly had culture conflicts and communication problems. Being separated by such a large distance also made working conditions difficult. Chiron, one of the largest and most successful biotechnology companies, was the only partner that Ciba conducted joint research with successfully and comfortably.

Ciba realized that it did not have skill, or could not tolerate, dealing with most researchers in the biotechnology area and decided to take another approach. In November 1994, Ciba paid \$2.1 billion to increase its stake in Chiron from 4% to 49.9%. The primary purpose of this deal was to improve Ciba's strategic position with the biotechnology industry, not necessarily just access to Chiron's knowledge base. Ciba intended to use Chiron as the intermediary to the rest of the biotechnology industry. Chiron had strong ties to other researchers at universities and other small biotechnology companies, and obviously possessed the absorptive capacity to interpret, understand, and apply these external knowledge sources. More importantly, they had the type of organizational culture that allowed them to deal with the biotechnology researchers. Without Chiron, Ciba had great difficulty trying to access these knowledge sources. This strategy of using an intermediary/interpreter proved very successful for Ciba. For example, in May 1995 Ciba-Geigy, Chiron and New York University collaborated on the development of a new approach for the optical mapping and sequencing of genes (Fisher 1995). In the deal Chiron interacts directly with NYU researchers and is granted an exclusive worldwide license to the technology. Ciba-Geigy receives a sublicense from Chiron to use the technology for its own research purposes.

The Ciba-Geigy/Chiron alliance seemed to work for a couple of reasons. First, Chiron effectively balances the need for a loosely structured, entrepreneurial culture with a clear sense of direction and focus. Their strong leadership fosters creativity, but only within certain boundaries - a situation described as "managed chaos" (Perry 1995). Second, Ciba-Geigy understood that for Chiron to be success-ful in its intermediary role, it had to remain autonomous and not be pressured to be controlled like a typical pharmaceutical company. Ciba-Geigy took a very "hands-off" approach and allowed the alli-ance to be loose and flexible. Both companies relied on mutual trust to develop a successful partnership. This must have been particularly hard for Ciba-Geigy to do after it already encountered so many dis-appointments with deals with other biotechnology companies. This example clearly illustrates how companies must be flexible in their management style during industry fusion to ensure that different knowledge areas can be integrated efficiently and effectively.

5. Conclusion

We have presented an argument about industry fusion, a phenome-non that leads to development of new industries through merging of technology and market conditions. At the heart of the industry fusion lies the development of some architectural innovations changing the boundary conditions of an industry. The important lesson for the industry fusion is that the rules of competition change as new guide-lines and standards of performance evolve. This blurs the distinction between competitors and a collaborator. In the telecommunication industry, for example, AT&T and Nokia are partners in the cellular telephone, but may be competitors in other areas. The traditional view of dealing with one's competitor on an adversarial fashion may not be prudent, as one may have to partner with the same competitor in the near future. This may seem to be confusing and may necessitate a certain level of strategic equivocality in dealing with one's competitors and collaborators.

Industry fusion dictates emphasis on creativity and ability to integrate different streams of knowledge in developing product and market opportunities. Rapid advances in information technology for example have created great opportunities in new services and products for many industries. We have provided some detailed information about the pharmaceutical industry integrating the advances in biotechnology. Integration of different knowledge streams requires a culture of organizational learning and development of internal technical capability to absorb external technologies.

Past strategies and procedures may create some hindrance. It is not easy to change the technological trajectories as firms develop their core competence in certain fields and consequently become entrenched in them. Firms have squandered away great opportunities as their past trajectories of growth and development did not match the new opportunities. Developing a strategic flexibility is more important in this dynamic environment.

The dilemma of specializing in certain core competence areas as opposed to broadening the technological base is a real problem. Here the management has to weigh the risks involved in the decisions and follow a course of action that will offer an acceptable level of risk while it would not foreclose the future opportunities. There are many mechanisms by which the risks may be shared among strategic partners and exploit the technology generated outside one's own organization. Outsourcing technology through various means, such as licensing, joint venture, acquisition, etc. may be effective means, if one implements them properly. Capability for organizational learning for innovation and creativity becomes the ultimate core competence to survive in such dynamic competitive environments.

Dynamics of technological competencies

Klaus Brockhoff

Dynamics of technological competencies

Klaus Brockhoff

1. Introduction

Different forms in which companies integrate technological knowledge into their production and transform it into marketable products can evidently exist successfully alongside each other. Economic explanations for this observation must therefore be sought in which a general failure of competition and permanent survival by defending monopolistic niches are not relevant in the long run. We suggest that economically relevant explanations arise from characteristics of technological progress which can be used to explain the wide variety of observations that can be made as companies adjust to such external conditions. Nevertheless, the explanatory model that is developed here is only a partial model.

2. Cost theory as a basis of explanation

The economic effects of technological progress vary widely and occur on various levels. A theoretical review can only be given in very brief form.

On a first level, costs of knowledge generation within one organization can be considered. The competitive effects of these can be devalued by radically new knowledge. On a second level, transaction costs become relevant (Picot 1993, c. 4194-4204). These result from asymmetric information and their possible opportunistic exploitation, which each partner in a process that is based on the division of labor strives to preserve. This causes the transaction costs which are a result of "frictions" that arise from the coordination between the companies participating in the generation of an output of goods or services. The level of these frictions and therefore also the amount of the costs is determined by cultural influences or value systems which are not specific to the transaction, as well as by the form of the information and communications systems. Transaction-specific influences depend on the specificity, the variability or uncertainty, and the frequency of the transactions (Picot, 1993, c. 4198 f.). These factors of influence (Brockhoff, 1992, p. 514-524), which can be affected not least by technological developments, are examined in more detail later.

In principle, it is assumed that transaction costs increase with increasing specificity, and that high specificity can be controlled less expensively through hierarchical coordination than through market-based exchanges. The transaction costs per transaction decrease with increased transaction frequency and thereby promote coordination via markets. Where specificity is low, increasing uncertainty will encourage coordination via markets; however, high specificity will incline more towards hierarchical coordination.

On a third level, we might observe opportunity costs. These arise, for instance, if binding relationships with particular partners inhibit more profitable relations with others. Within networks of companies, such as suppliers and users of jointly developed new technologies, such opportunity costs are of special importance.

What now needs to be clarified is: how does new technological knowledge affect these costs, and thereby favor different types of organization for its use in new products? For this purpose we need first of all to refer to some specific characteristics of knowledge.

Knowledge can be used any number of times without being exhausted. This explains the interest in its dissemination from the point of view of every other potential user of the knowledge, because dissemination helps to reduce specificity. The specificity can be maintained through non-dissemination and thereby becomes a basis for the attainment of "rents".

Knowledge can be combined with other knowledge, thus creating new technologies that help to satisfy a wider range of needs. This reduces uncertainty about the exploitation of the original knowledge base. It leads to "architectural innovations": "Existing technologies are applied or combined to create novel products or services, or new applications. Competition is based on serving specific market niches and on close relations with customers" (Tidd, Bessant, Pavitt 1997, p. 165). Two examples will demonstrate this. *The garden tool manufacturer Gardena views knowledge as a condition for a high level of innovativeness and combines it with high vertical integration to ensure delivery capability in the case of high seasonal fluctuations, or uncertainty of demand* (FAZ Sept. 18, 1996, p. 26). *Henkel developed its core competence in the adhesives industry from the packaging needs that were not adequately met by its suppliers in the 1920s* (FAZ Jan. 9, 1998).

Where there is cumulative technological progress, knowledge owners can achieve synergism from a technological point of view from the integration of new elements of knowledge. Whether these can be exploited in the market is a different question, which depends in particular on whether this market remains homogenous or will itself be heterogenized with the increase in offerings based on the new technologies. This can lead to excessively high costs if all needs are satisfied from a single source.

These characteristics of technological knowledge are important for the following considerations.

3. Developing a "knowledge spectrum"

3.1 Dissemination of knowledge

At one extreme, technological knowledge can be monopolized by one person or institution. Alternatively, it might be disseminated throughout a very large number of people or institutions. The owner of monopolized knowledge will rarely have an interest in its dissemination (network products could form an exception to this); however, s/he can hardly effectively resist it in the long term. It is demonstrated that the monopolization of knowledge stimulates "technology races", while when knowledge is disseminated ("technical parity competition") other variables determine the competitive position of companies (Miller 1995, p. 511-524). This suggests instability of a position on an axis depicting dissemination of knowledge. The transition from monopolized to disseminated knowledge appears to be influenced by the type of technological progress, amongst other things. Empirical technological progress is more likely to enable monopolized knowledge, whilst cumulative technological progress makes wide dissemination easier through the use of the underlying theory. This character of the technological progress can change in the course of time as theories are developed to explain and forecast effects of new knowledge. *For example, two consecu-*

tive phases are described during the development of the radio in the history of technology: "the emergence of a pre-technical system, characterized by a high degree of empiricism and the concentration of innovators on solving operational problems arising from the incapacity of the dominant technical system to support further development of new services. When these tensions decreased, and when scientific knowledge was able to follow, and even precede technical evolution, the pre-technical system was able to generate by itself new services or products"(Griset 1995, p.47).

3.2 Complexity of output

Knowledge can be employed in the form of very simple products or very complex products. The combinatorial properties of technological knowledge can support a development towards more complexity. The transition from one level of complexity to another is influenced by the latent or manifest demands of buyers on the one hand. On the other hand, technologically complex products, in particular, place high demands on learning within companies (Griset 1995). It is not certain that all competitors in an industry can meet these demands, as they tend to increase costs of production. *Griset argues that the rise of electronics in the "radio industry" required the mastery of very complex technological systems with elements of hardware and software, which RCA mastered, while the "classical" companies within the industry such as Marconi, Telefunken, or CSF did not* (Griset 1995, p. 54 ff.).

Let us now examine complexity somewhat closer. The creation of complex means to satisfy needs is not one-dimensional. On the one hand, complexity can arise through the integration of more parts in one product (Henderson, Clark 1990, p. 9-30); on the other hand, however, it can also arise through the delivery of an increasing number of product functions, which can only occur as a result of a change in the range of performance of the same number of product components. These complexity dimensions are in principle independent of one another (Singh 1997, p. 6), so that they can appear in every possible combination. Similar ideas on complexity have been expressed many times. Sometimes, these ideas have been empirically substantiated (Henderson, Clark 1990, p. 9-30; Fujimoto 1993, p. 165 ff.).

The ability to generate synergism from combinatory use of knowledge could lead to a tendency to develop ever more complex products, unless this is balanced by a cost factor such as the learning mentioned above. This indicates another instability on the complexity spectrum.

3.3 Knowledge spectrum

Dissemination and complexity represent scales which span what we call a knowledge spectrum. Each scale can be divided into any desired level of fineness. Only four particularly prominent points on the scales are considered here. This makes it possible to keep the representation within manageable limits. The four positions shown by combining selected points on the two scales are dealt with in succession. These positions are illustrated in Figure 1.

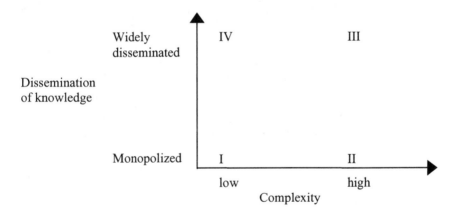

Figure 1: The knowledge spectrum

Often, only one of the quadrants of this diagram is dealt with. This creates the impression that only this position secures corporate competitiveness. The theory that such particular positions represent absorbing states, towards which all successful companies must develop, carries this idea further. In one article it is said: "Industry will gravitate toward the concept of core competencies, or strategic technologies, to focus its efforts" (Bridenbaugh 1996, p. 162). This seems to imply that when this state is reached no further changes will follow. Such theories overlook the dependence of successful management on adjustments to economic or technological constraints, which in turn may be subject to radical change. It is therefore not justifiable to recommend particular positions to all companies, as has recently occurred when concentration on core competencies or R&D cooperation as its consequence or the use of system suppliers have been suggested as particularly successful strategies.

4. Four quadrants in the knowledge spectrum

4.1 Monopolized knowledge in non-complex products (I)

At the extreme, one individual can create or possess new knowledge which can be exploited economically, even if it is in the form of a product which is not complex initially. This product can be a radical innovation. The owner of the new knowledge simultaneously creates information asymmetries vis-a-vis other people. *The Benz motor car, for which petrol had to be bought at the pharmacy and which was driven on traditional dirt roads, the crystal receiver with the headphones for listening to radio transmissions, and the first Apple personal computer are examples of this.* At this stage, the products do not require complementary products for their performance. They appear as "stand alone products". The specificity is high, as is the uncertainty of their future success, and only a few transactions are expected at first. Production takes place by hierarchical coordination.

The technological progress which makes such products possible can be "radical", in that it enables considerable increases in performance compared to traditional products. This makes the new products potentially desirable by customers. However, this technological progress can also be "disruptive": an innovation which opens up previously undeveloped market segments, and enables value added there, even if a majority of customers detects underperformance with respect to characteristics presently held in high esteem (Christensen 1997, p. 15 ff.). Established market players can easily overlook this kind of technological progress, because it appeals only to a small customer group, possibly allows only for lower margins at first, and does not (yet) meet the expectations of the mainstream customers. Should the major suppliers recognize this type of technological progress, they might choose to push ahead the further development of the old technology because by doing so they can protect their sunk cost from the pending devaluation of the technological potentials tied up in it. The "sailing ship effect" describes this response. It is bound to fail when the performance potential of the new technology is significantly higher than that of the old technology.

The hierarchical coordination in the first quadrant of the knowledge spectrum does not remain stable. There are various reasons for this. Firstly, the integration of knowledge in a product does, as assumed here, lead to a "feasible" solution, but there may be superior technologies which can fulfill the same purposes. In case of a more or less simultaneous perception of needs, in particular, this can trigger demand pull technical progress, which is characterized by the fact that the same outputs can be achieved by very different factor inputs. As a rule, different factor inputs cause different costs; the most cost-effective combination of factors will be superior, all other things being equal. This is indicated by a multitude of parallel inventions documented in the history of technology (Lamb, Eaton 1984, p. 47 ff.) as well as the attempts to secure specificity through comprehensive fencing of important patents by their applicants (Spero 1990, pp. 58-67). Both of these are indications of the technical races referred to above. The existing specificity can, therefore, get lost. *An interesting indication of this problem occurs in the computing industry: "When Apple had been the only game in town, the engineering team could design a machine to please themselves. But competition, and the need for larger sales volumes, raised the stakes and restricted design freedom"* (Penzias 1989, p. 185).

Secondly, a monopolistic technological position cannot necessarily justify economic specificity, if it is in a firm's best interest to give it up. The technological knowledge is made generally accessible in order to gain an economically more advantageous position through the initiation of complementary products (e.g., software for a particular hardware), the coverage of markets which cannot be easily accessed by the firm itself (e.g., through licensing), to secure future participation in a cumulative technological advance, or to obtain network effects.

Thirdly, a legally protected unique technological position with specificity cannot usually be maintained in the long run, because the period of legal protection is limited. Specificity can be lost through the appearance of imitators who are attracted by the rents created by that specificity. *The rise of generics manufacturers on the ethical pharmaceutical markets is a good example of this.*

Even when the period of protection is not limited, technological progress and the differentiation of demand can mean that the monopoly must be surrendered. *In telecommunications, it has been observed that technological progress "...enables a wide variety of outputs and therefore leads to a diversification of the markets which can no longer be managed by a state administration. Here, therefore, a causality is formulated which posits technological development as an action component, and the economic opening of the market as its consequence"* (Witte 1997, p. 8). This "causality" is also supported empirically (Witte 1997, p. 13). The monopoly cannot prevent the appearance of technologically related outputs, which in the subsequent period develop into technological alternatives to the core output. This suggests that market heterogeneity makes the exploitation of technological synergies by one market player seem uneconomic.

Fourthly, radically new knowledge, in particular, makes its owner into an interesting acquisition target or into a highly esteemed cooperation partner. If this occurs, however, knowledge is transferred to others. Acquisition of the whole firm is more expensive than the acquisition only of the desired technology or cooperation for two reasons: firstly, a higher degree of control is obtained, in particular over the owners of the knowledge, and this has to be paid for; secondly, it is not always possible to avoid the acquisition of assets that cannot be used optimally by the acquirer along with the acquired firm.

In the case of cooperations, the hard-to-come-by knowledge can be so well protected that its owner can put a visible and economically valuable stamp on the cooperation. *The reference to the label "Intel inside" on many PCs produced by different manufacturers documents this. But it is not an entirely new phenomenon. Around 1920, L'Hohlwein cigarettes were sold. The designer of the boxes, Ludwig Hohlwein, granted a license to the manufacturer, the Menes cigarette factory in Wiesbaden, to use his name as a trademark* (exhibition catalog Altona Museum 1996).

Thus, the first position in the knowledge spectrum is not stable. A firm which takes this position may either leave it for self-interest or will find itself confronted by attacks on its position from others.

4.2 Monopolized knowledge in complex products (II)

Knowledge can be "combined" with other knowledge without being exhausted. This creates new opportunities for the satisfaction of needs. These combinations increase the complexity of the product which is to satisfy the wants of potential buyers. The possibility that the demands of the customer may be exceeded and that undesirably complex products could be created as a result will only be mentioned in passing here. It is a result of incomplete information. This "overengineering" is a first indication of the instability of the economic position adopted.

There are many examples describing the second position in the knowledge spectrum discussed here. We are told that: "for years, Eastman Kodak was the only company that did all three: cameras, film, and developing" (Design Management Institute 1993, p. 5*), thus offering a very complex product. In another case "plain old telephone services (pots)" became over time "progressively combined forms of communication..., which include speech, text, data, stills, and eventually also moving pictures"* (Witte 1997, p. 8). *Initially these opportunities were bundled by a monopoly, however, they then broke the bounds of this market regime with their increasing breadth and complexity. Another case: The Meissen china manufacture even today is proud of excavating its own clay, developing and mixing its own colors etc. All of these are made into complex works of art.*

The treatment of knowledge that leads to complex products is an important part of the training of engineers and scientists. *Pavitt points out that research is carried out "precisely to train technological problem-solvers to integrate knowledge from a variety of disciplines in the development and use of complex systems"* (Pavitt 1993, p. 133). *Kesselring wrote a "theory of design (composition)", which was intended to support the creative activities of an engineer in the interaction of invention, design, and formation with regard to economic, ethical and other goals* (Kesselring 1954). *The principle of variants, that is, the search for similar solutions to meet a need which is formulated as a task, and the principle of generation, that is, searching for existing and re-combinable assemblies or the development of new assemblies and components, are nowadays taught as elements of technical problem-solving* (Eversheim 1996, pp. 7-20 ff.). *It is important also to mention various creativity techniques, in particular morphological analysis, which Hauschildt describes as "structuring concepts for the generation of alternatives"* (Hauschildt 1997, p. 311 ff.). *In the above sense, these therefore support the principle of generation.*

Because of the multi-dimensionality of "complexity" referred to above, sustaining competitiveness requires the pursuit of both the technological development of the "product architecture" and the performance of the individual elements or modules of a product which are connected in the architecture of the product.

In particular, when the interfaces between the components or the causes of the improvement in their performance are difficult to ascertain from the outside, this again forms a basis for specificity. In so far as the increase in complexity also entails coverage of shrinking market segments, down to one-piece production, the number of transactions decreases (Tidd 1997, p. 6 ff.). So does the number of possible transactions where external sourcing of components is attempted. For both these reasons, it is therefore to be expected that with growing complexity an increasing share of the value added of the product will be generated within the firm.

The resulting coordination costs are seen as one of the main obstacles to economic efficiency. For single-product firms, Gutenberg did not consider coordination costs (Robinson 1936, p. 87) as being a factor limiting the size of the firm "under any conditions and in any magnitudes likely to occur in practice" (Gutenberg 1956, p. 35). However, such an effect cannot entirely be ruled out and - which appears more crucial here - the argument does not take into account the case of variable complexity or multiple types of products. *An example of the shift of cost levels is the success factor for book clubs. In 1948 it consisted of a close link to a printer, which could deliver the desired titles on time and promised additional incremental profits. Forty years later printing has become a commodity which is offered on a competitive basis. Thus, efficient book clubs do no longer operate their own print shops. Another example: "As Apple grew ... so did the 'team'. Specialization became necessary, and with it a need for more coordination. Now teams split off ... and some tasks ... got done twice while others ... went neglected"* (Penzias 1989, p. 183).

Tidd suggests that firms which find themselves in the situation described above gear their communication and information relationships too much to the performance of the modules and assume that the traditional architectural knowledge will persist (Tidd 1997, p. 7); they can then be forced out of their competitive position by new architectural ideas. Research and development activities in successful firms will therefore extend to both aspects of complexity. However, the appearance of disruptive technologies and the manufacturer's disregard for them (on the grounds already stated with respect to the first quadrant and in the field of technology, in particular, because of the existence of the well-known "not invented here syndrome" (Katz, Allen 1982, pp. 7-19) attack the position achieved. A necessary condition for recognizing radical and disruptive innovations can be met by investment in research, if this is used to build the potentials for identifying and absorbing external knowledge (Brockhoff 1997).

The creation of a special organizational unit for the development of new technology and its later exploitation is advised (Christensen 1997, p. 20, 101 ff., 197 ff.), in order not to lose the opportunity for innovation by being too strongly bound to the past. This unit will tend to move the firm back towards quadrant I, although on a different technological level from that which the firm originally held.

A second reason for instability once more lies in the fact that knowledge cannot generally be completely protected in the long term. In particular, knowledge about the technical interfaces between modules (and therefore the basis for architectural innovations) would, as manufacturing becomes increasingly standardized (thus creating market entry barriers through economies of scale and scope), have to be documented or passed on to many owners of knowledge. This increases the chances of unintentional leakage of this knowledge, especially since, in the case of successful activity, systematic efforts will be directed at the absorption of this knowledge by external competitors (Lange 1994).

A third reason for the instability of the situation described here could be that, despite the fact that the knowledge components to be combined can be identified unambiguously and cost-effectively, their in-house development for productive exploitation is associated with higher costs than development or delivery from outside the firm (Brockhoff 1997). *An example can be cited here, too: The search for a supplier for sodium perborate, which Henkel needs for the production of its leading brand detergent Persil, led to Degussa AG. In order to secure the supply Henkel held an indirect share in Degussa AG until 1997*[1].

This shows in a particularly striking way that firms have difficulties to stay in position II of the knowledge spectrum. Moving towards position III they will aim to secure relationships as long as the knowledge has to become common and shared by many. If a precise description of the object of a transfer is not possible, as in the case of necessary research and development, then instead of a contractual obligation it is advisable to attempt to build trust-based relationships (Ring, van de Ven 1994, p. 93). This is especially true in the development of complex systems which have parts which interact with each other, and therefore do not allow exact prior formulation of the final technical interfaces or their characteristics. Trust is "produced through interpersonal interactions that lead to social-psychological bonds of mutual norms, sentiments, and friendships" and is supported by the motivation "to seek both equity and efficiency outcomes because of the desire to preserve a reputation for fair dealing that will enable ... to continue to exchange transaction-specific investments under conditions of high uncertainty" (Ring, van de Ven 1994, pp. 93-94). These descriptive statements have, however, only rarely been analyzed in time-series studies of development for real situations which are characterized by partners changing their employers quickly, where there is a high level of environmental variability, where inflexibilities in the labor market exist, etc. Perhaps the securing of stability among partners through trust building is not quite as effective as supposed (i.e., it causes relatively high costs), because one observes that when discontinuous technological change occurs, contractually secured forms of cooperation (joint ventures, technology licensing agreements, and various forms of R&D consortia) are preferred to generate knowledge (Lambe, Spekman 1997, pp. 102-116). Trust building might need to be supplemented by formal legal agreements (which could then be interpreted as a cost-saving measure). *This is shown by the ABS case: in 1965, Daimler-Benz AG hinted to Robert Bosch GmbH that it should address itself to the development of an anti-locking system* (Bingmann 1994, p.

---

[1] The GFC Gesellschaft für Chemiewerte mbH holds more than 25% of Degussa AG; Henkel KGaA had a holding of 46 % in GFCmbH, the remaining shares being held by Dresdner Bank AG and the Münchner Rückversicherungs-Gesellschaft.

780). *It also supported Bosch's acquisition of Teldix in 1973, where corresponding developments were already well advanced* (Bingmann 1994, p. 786 f.). *Intensive discussions took place between Daimler-Benz and Bosch in "weekly meetings", but without the conclusion of a binding development contract or marketing contract. Although Bosch had agreed to an - undefined - lead time for the use of ABS by Daimler-Benz, in 1977 the company offered the development to other auto manufacturers as well at a stage when it was practically functional and ready for scale production. One week before the planned press presentation by Daimler-Benz, Bayerische Motorenwerke (BMW), aware of Daimler-Benz's timing, launched a press release with its own announcement of ABS. "The breach of the decades-old 'Good Faith' process led to a substantial loss of trust in Robert Bosch GmbH on the part of Daimler-Benz, which was still being referred to as 'ABS-Trauma' in 1988"* (Bingmann 1994, pp. 789-790). *It seems that Bosch gambled its trust capital because by that time competitors were hot on its heels with their own developments, and rapid establishment of its own development in the market to amortize the development expenses could not be guaranteed if it only supplied one firm.*

Assuming that Bosch's behavior was rational, this can be explained by considering opportunity costs. Obviously, the sum of future transaction costs between Bosch and Daimler-Benz as well as the opportunity costs of delayed information and supply to other automotive manufacturers if the company had been committed to this partnership for a certain period were higher than the comparative costs. These are made up of the (higher) transaction costs arising from supplying to several auto manufacturers as well as the opportunity costs for lost follow-up business. The situation is represented in Figure 2.

If we are to assume a decrease in the transaction costs with increasing length of a relationship (because of the reduction in uncertainty as well as the larger number of transactions), then cooperating firms would have to continue their commitment or replace it with even more favorable market conditions. The departure from the adopted development direction can be explained by taking opportunity costs into account in addition to the transaction costs.

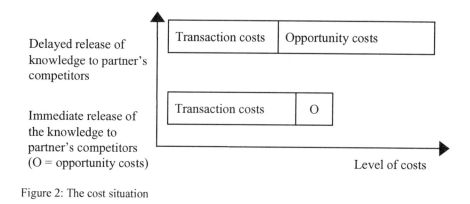

Delayed release of
knowledge to partner's
competitors

Immediate release of
the knowledge to
partner's competitors
(O = opportunity costs)

Figure 2: The cost situation

In conclusion, we find that a position in quadrant II is not stable either. It is technologically vulnerable and changeable. The economic consequences can then force abandonment of the position.

4.3 Disseminated knowledge in complex products (III)

In digression from position II we posit that the knowledge required for manufacturing product modules is no longer monopolized. If we assume that efficiently designing and bridging interfaces between modules of complex products can be a special technological competence, then in the case of small sales volumes of products which may even be constructed according to the individual demands of the purchaser, collaboration between several firms in production may be cost-efficient. Tidd has formulated the thesis: "Technological and market complexity is positively associated with network participation" (Tidd 1997, p. 4). In such a case, the most economic solution to satisfying the customer can be, for example, the formation of a working group or a network of firms with a lead firm. Depending on the design of the relationships, different forms of hybrid organizations of networks of firms are created.

Network organizations exploit specialization and thus information asymmetries. In particular for the lead firm, one assurance against the opportunistic exploitation of information asymmetry in such organizations could be to select a broader range of research and development activities than what is required by production. In addition to the knowledge which is necessary for internal activities, knowledge should be provided by research and development that is necessary for the identification of technological alternatives for the externally produced modules. *High transaction costs were incurred by Hoechst AG until 1925, when it tried to rely on external knowledge and cooperations. These costs resulted from profit-sharing agreements with the external possessors of knowledge, and they were greater than the costs of internal personnel. This led to the company's establishing Hoechst's own R&D laboratory. When some of the competencies later had to be transferred to IG Farben laboratory in Leverkusen this was again seen as a disadvantage* (Wimmer 1994, p. 149 ff.). In other words: at least a partial reduction in information asymmetry is viewed as economically reasonable, in order to limit the temptation of partners to behave opportunistically. Therefore, the degree of specialization in research and development should be lower than that in manufacturing. Different degrees of specialization are not uncommon between the marketing departments and production departments of system manufacturers. The same idea is here transferred onto a further functional area.

The situation described here may not be stable either. On the one hand, deviations are to be expected because, as a result of increasing sales volumes, standardization of the modules and their interfaces becomes possible and necessary. As a result, special skills in interface structuring might get lost. On the other hand, such standardization may also enable an increase in sales. It is to be expected that for both these reasons the transaction costs will decline, so that the situation described in quadrant IV may arise.

It is also noteworthy that the function of the lead firm can get lost through particular efforts by the supplier in the innovation process. Direct suppliers and system suppliers in the automotive industry can be differentiated by, amongst other things, the fact that the latter show higher product innovation skills, patentability, and readiness for know-how transfer in the research and development processes. At the same time, however, they are subject to less intervention by the car manufacturers in their research and development and their process innovations (Gaitanides 1997, p. 750). They therefore try to counter the dissemination of knowledge by monopolizing their own knowledge. If we take into consideration that the system suppliers earn higher returns than the direct suppliers (Gaitanides 1997, p. 741), then this is indicative of a stronger position vis-à-vis auto manufacturers. The originally leading role of the auto manufacturer is, correspondingly, weakened in relative terms. This, in turn, can be an incentive to standardize the interfaces, in order to place suppliers under competitive pressure.

4.4 Disseminated knowledge in non-complex products (IV)

As a consequence of standardized interfaces, several firms will offer modules in competition as the knowledge required for each module is increasingly disseminated. The definition of the interfaces makes division of labor in invention easier. Then, it will no longer be only the system leader who combines the modules for his customers: the customers themselves will put the modules together from complementary parts. In comparison to the total bundle of outputs which arises as a result of this, each module is less complex. *Examples are tuner, receiver, tape drive and speakers instead of a "radio" or a truck with different tires, units and chassis all made by different manufacturers at the customer's order. In the computer industry, it was found that: "After thirty years of progress at a rate of 25 percent per year and after internationalization of the know-how required, many of the hardware technologies in computing and telecommunications are available as commodities. Therefore, these technologies in themselves are no longer the source of comparative advantage that they once were"* (Armstrong 1996, p. 152). *Elsewhere it is stated: "Rather than every computer company in the world designing and building its own proprietary parts, independent suppliers provided standard building blocks. As a consequence, a wide array of companies could build machines that performed across a broad spectrum of price and performance features"* (Yoffie, Pearson 1991, p. 4). If the technological interface problems cannot be resolved by a manufacturer which assembles the elements, this can result in serious competitive disadvantages from the customer's point of view (Yoffie, Pearson 1991, p. 6). If the problems are solved in a way which is visible to the outside world, however, competitors are given the opportunity to imitate the output, *e.g., in the computer market, the "IBM clones"*. If standardization occurs, competitive advantages can no longer be derived from technology developed in-house, but must be obtained from other outputs (e.g., services, consultancy, reliability, upwards compatibility). The exploitation of technologically based advantages might be shifted to the suppliers of the modules (*"Intel inside" is then put on the computers*), who will take a position within their markets in accordance with quadrants I or II as described above. In such a situation, it appears extraordinarily difficult to regain the position of the technology-developing lead firm once the role of "assembler" has been accepted. *The example of the failure of the development of the Apple Lisa computer demonstrates this. In contrast to developments which were immediately previous to this (but not Macintoshes), Apple itself wanted in this case to be involved again in the programming of the operating system* (Swanger, Maidique 1985, p. 18). If a supplier cannot be quickly changed, its problems in maintaining or advancing a desired state of the art are naturally also problems for its customer, the "assembler". *Delays in the supply of Motorola chips in 1990 obviously did significant damage to Apple's image as the performance leader in personal computer* (Yoffie, Cohn, Levy 1992, p. 10). *IBM, in contrast, kept on its own microchip production for internal use and secured its supplies from Intel by investing in the company* (Coleman 1987, p. 1). *Beyschlag produces 5 billion resistors per year for the automotive and the electronics industries. It relies on proprietary production processes, reliability and product quality, and engagement in problem solving together with its customers.* An alternative for the powerful customer to secure its competitive position is to demand that the supplier must ensure that its products are interchangeable with others (Bingmann 1994, p. 800).

The manufacturers of the individual modules which are standardized at their interfaces will aim for performance improvements in the modules only if the customer can thereby improve the total performance of its system and wishes to do so. Otherwise, the realization of competitive advantages will focus on cost reductions prepared by process development, which as far as necessary will be passed on in the form of lower prices. Similar considerations also apply to the case where a dominant design has become established (Utterback 1994).

The stability of a situation shown in quadrant IV can also be disturbed by radical innovations, which in turn can lead to quadrants I or III, but at a new technological level and therefore, probably, with new market players.

Profit-maximizing module manufacturers or system suppliers may perceive their own situation as offering room for improvement and act accordingly. *When the planned establishment of a development laboratory in Germany by an American automotive industry system supplier was announced, whereby the "required closeness to the auto manufacturers" was to be achieved, acquisitions were also promised. Two reasons were given for this: firstly, obtaining more customers; secondly, reducing the dependence on individual customers* (FAZ Jul. 19, 1997, p. 16). If this strategy makes the module supplier stronger, that is, makes it possible for the supplier to assert higher prices because control by competition is reduced, a firm may arrive once more at the point of technological backward integration of sub-contracted parts. Such a firm would be aiming to change its position from IV to II or even I for economic reasons.

5. Summary and further considerations

(1) It has been demonstrated that the situations considered in the four quadrants of the knowledge spectrum are not stable, but are vulnerable to technological innovations, amongst other things. The positions represent ideal types at prominent points on a continuum of knowledge dissemination and combinatorial knowledge use for product complexity. Some of the transitions between III and IV have already been described in more detail (Meyer 1994, p. 102 ff., 164). Because of the technology-related vulnerability of every situation described, it cannot be expected that all firms or even business units will develop towards a single position, as is sometimes maintained. An overview of the destabilizing influences discussed here is given in Figure 3.

(2) Broad knowledge dissemination and high complexity, as considered here, do not represent "natural" extremes of the respective scales. The dissemination of knowledge could in theory be extended to every person. This, however, is not practicable on economic grounds, amongst others. The maximum level of dissemination is determined by learning costs, storage costs of the knowledge, and the value of knowledge. Complexity could also increase further from any attained level. From an economic point of view, this is chiefly to be expected when the costs of coordination of the complexity-inducing elements of a product decrease. The development of electronic communications media could, in this sense, offer an opportunity for an increase in complexity. Whether this condition is then exploited in one firm or whether it leads to new forms of cooperation between firms is partly determined, in turn, by the dissemination of knowledge.

I: Lower costs because of competing technology (parallel invention)
Loss of specificity in the case of network effects
High costs through differentiated technological progress and differentiated demand
Being overtaken by radically new knowledge

II: High coordination costs and neglect of innovations in the case of "architectural knowledge"
No ability to protect knowledge permanently
High costs of internal generation of knowledge
High opportunity costs in the case of trust-based commitments to particular customers as development partners

III: Opportunistic exploitation of information asymmetry
High cost of securing "minimum knowledge" outside the range of core products
Standardization of technological interfaces

IV: Costs of securing a "minimum knowledge" for the "assemblers" of the parts
Being overtaken by radically new knowledge
Monopolization by the module manufacturers

Figure 3: Destabilizing influences on the positions in the spectrum of knowledge

(3) Furthermore, it is interesting to observe that the direction and scope of research and development are neither confined to the range of activities of other functional areas of the firm, nor are they the same for all ideal types considered. This is significant for technology management of this function and general management. In reality, the more diversified a firm is, the harder it is to perceive this significance.

(4) Each business unit of a firm can find itself in one or another of the four quadrants and be attacked more or less strongly. Different research and development policies should be chosen accordingly. The considerations expressed here may also help to explain the trend observed in the last few years towards the formation of independently operating business units or the spin-off of independent companies from a firm.

(5) Finally, it is apparent that the transition between the situations described does not take place according to a mechanistic, sequential pattern. The advantageous transitions are rather determined by eco-

nomic conditions which are in part technology-based and which can be formulated using the concepts of transaction costs and opportunity costs. If we posit that knowledge is not lost (which, however, can occur in individual cases, as shown by, e.g., the repeated reinvention of the process for manufacturing ruby glass), then transitions from III or IV to I and II with the same knowledge are not primarily explained by technology, but by economics. All other transitions can have both technological and economic grounds.

The view developed here digresses from those models that make plausible very specific successions of innovative activities. One example for such models is the Abernathy-Utterback model of product and process innovations (Utterback 1994). This model indicates movements from our field I towards higher complexity, particularly in the phase where process innovations become relatively more frequent than product innovations. Eventually, more disseminated information might also be observed in the model. However, we do not assume a definite phasing of such innovation types.

(6) To date, no studies are known which empirically describe the dynamics of these transition processes in terms of the factors of influence considered here.

A study of firms which offer complex systems (e.g., radar systems), that is, firms which could be positioned either in situations II or III as described above, or - for individual modules - in situation IV, offers initial indications (Paganetto et al. 1998). The authors consider four "structuring alternatives"[2]. However, only four of the 16 possible transitions between these (observing two different points in time) are considered [3]. Particular attention is paid here on the one hand to development capacities and their ability to deliver the ne-

---

[2] These are: (a) development internal, production internal; (b) development internal, production external, assembly internal; (c) development external, production external, some in-house development capacity; (d) development external, production external, no in-house development capacity.

[3] The description is not quite clear. As far as can be ascertained, the following cases are considered: "persistence of vertical integration" from (a) to (a), whereby the structural alternatives in the footnote above are called by the corresponding letters; "de-verticalization" from (a) to (b), (c) or (d); "partial de-verticalization" from (b) to (c) or (d); "vertical integration" from (c) or (d) to (a) or (b); "re-verticalization", which corresponds to the last case, but refers back to a different case history.

cessary level of technological progress cost-effectively (in particular to pioneer or be a follower in radical innovations). On the other hand, the production capacities for modules are considered, as well as the capability of realizing economies of scale (which does not promise crucial advantages in markets with a very small demand), of avoiding supply bottlenecks for critical resources or of keeping the relative factor costs low, and defining the technological interfaces between modules. One of the important conclusions drawn from the observation of the transitions is that an "appropriate level of control over each technology" is necessary. This means, therefore, that even in the case of cooperations or the procurement of modules from outside a firm, the firm should maintain in-house the capability to judge technological progress competently as well as access to its results. This agrees with the idea considered above which posits that the breadth of the development tasks does not necessarily have to match the breadth of the production tasks. It appears that this necessity has not been considered in every one of the core competency-based reorganizations of the past years. Disadvantages arising from this are not visible immediately, but only after a time lag. They can, however, only be corrected at that stage by internal efforts which also have a delayed effect, or by the expensive acquisition of external knowledge, e.g., by acquiring a company. This is one of the dangers of neglecting the dynamics of technological competencies.

In our presentation, information drawn from reports and case studies has been used to support the reasoning. There has been no comprehensive empirical test of the main question: How do economically relevant characteristics of technological progress interact with a form of corporate organization designed to secure competitiveness and with markets? This question offers opportunities for further research.

Organizational adaptation and innovation:
The dynamics of adopting innovation types

Fariborz Damanpour / Shanthi Gopalakrishnan

Organizational adaptation and innovation:
The dynamics of adopting innovation types   **57**

Organizational adaptation and innovation: The dynamics of adopting innovation types

Fariborz Damanpour / Shanthi Gopalakrishnan

1. Introduction

Innovation is regarded as a focal point of an organization's strategy and a crucial element for its long-term strength and survival (Tushman, Anderson 1997). Organizations adopt innovations to introduce changes in their outcomes, structures, and processes in order to maintain or improve their level of performance or effectiveness. These changes can be the direct result of managerial choice or can be imposed by external conditions. For instance, a performance gap perceived by top executives of an organization creates a need for change, which in turn stimulates the adoption of innovation to reduce the gap (Zaltman, Duncan, Holbek 1973). Similarly, environmental change or uncertainty creates a need for change in the strategy and/or structure of an organization. This, in turn, provides the impetus for the introduction of innovations (Damanpour, Evan 1984). Regardless of the origin of organizational change (internal or external), innovation is a means of creating change to ensure adaptive behavior.

Because of its important role for organizational adaptation, its perceived effectiveness, and its contribution to organizational performance, innovation has been studied widely by researchers in a variety of fields. A dominant feature of these studies, however, has been the inconsistency of their findings. Downs and Mohr (1976), in their critical review of innovation research, stated that extreme variances have occurred regularly among the findings of innovation studies, and that the variation of results is beyond interpretation, and findings have not been cumulative. In a more recent review, Wolfe (1994) expressed the same concern and stated that, perhaps, the most consistent result of innovation studies is that the results are inconsistent. In an attempt to remedy the instability in the results of innovation research, researchers have developed intermediate or subtheories of organizational innovation. The majority of these theories are based on differentiation between types of innovation.

The distinction between innovation types is considered necessary because neither do all types of innovation have identical attributes, nor do they relate equally to the same predictor variables (Damanpour 1987). Further, the process of adoption or implementation of different types of innovation in organizations is not identical, some follow a top-down process, others a bottom-up process (Daft 1978). Thus, a study of one type might produce results different from a study of another type (Daft, Becker 1978). Researchers have primarily examined differences between administrative and technical innovations (Daft 1978; Damanpour 1987; Kimberly, Evanisko 1981), product and process innovations (Barras 1986; Damanpour, Gopalakrishnan, 1998b; Utterback, Abernathy 1975), and radical and incremental innovations (Dewar, Dutton 1986; Ettlie, Bridges, O'Keefe 1984; Nord, Tucker 1987). In this paper, we review these studies and present their findings.

Contrary to the assertions of differences between innovation types in the intermediate theories of organizational innovation, meta-analytic reviews demonstrated that statistically significant differences do not exist between the determinants of innovation types (Damanpour 1991, 1992). This lack of empirical support for differences between innovation types calls for a different conceptualization of the role of innovation types in organizations.

In this paper, we emphasize congruency in adoption over differences between types. We posit that, because innovation is a means of helping organizations adapt to external and internal changes, congruency in the adoption of innovation types is crucial to organizational effectiveness. Organizations need to manage the adoption of innovation successfully over time in order to survive, compete, and prosper in the technological, commercial, and social environments. Changes in the organization's strategies, structure, and processes affect various organizational systems. A change in one of the organizational systems would produce a compensatory change in the other systems. A fit among the internal systems would favor the effective operation of the organization. To influence performance positively, therefore, different types of innovations should be introduced concurrently so that the organizational systems remain in balance and mutually reinforce one another.

2. A review of theories of innovation types

2.1 Administrative and technical innovations

Among typologies of innovation, technical and administrative innovations have most widely been distinguished in studies of organizational innovation. *Technical innovations* pertain to products, services, and production process technology; that is, they can be the adoption of an idea for a new product or a new service or the introduction of a new element in an organization's production process or service operation. Technical innovations are those related to the primary work activity of the organization and can be either product or process innovations (Daft 1978; Damanpour, Evan 1984; Knight 1967). *Administrative innovations* involve organizational structure and administrative processes; that is, they can be the adoption of new ways to recruit personnel, allocate resources, give rewards, and structure tasks or units. Administrative innovations are indirectly related to the primary work activity of the organization and more directly related to its management (Daft 1978; Damanpour, Evan 1984; Kimberly, Evanisko 1981) and are largely process innovations.

2.1.1 Organizational lag model

Evan (1966) defined organizational lag as the discrepancy in the rate at which new technical and administrative ideas and behaviors are introduced in organizations. Following Ogburn's (1922) hypothesis of "cultural lag", which pertains to the societal level rather than organizational level of analysis, Evan advanced an analogous hypothesis that administrative innovations in organizations tend to lag technical innovations.

Damanpour and Evan (1984), applying Rogers' (1995) proposed relationships between the five attributes of innovations and the rate of adoption of innovations, provided theoretical support for Evan's hypothesis. Rogers proposed that relative advantage, compatibility, trialability, and observability of innovations were positively related to the adoption of innovations, while complexity was negatively related to the adoption of innovations. Technical innovations are more observable, have higher trialability, and are perceived to be relatively more advantageous than administrative innovations, while administrative innovations are perceived to be more complex than technical innovations to implement (Damanpour, Evan 1984). Furthermore, in an empirical study of the adoption of innovations in libraries, Damanpour and Evan found libraries in all size categories adopt technical innovations at a faster rate than administrative innovations (Table 1). Swanson (1994) has applied organizational lag model to the adoption of information systems innovations in organizations, however, his theory has not yet empirically been tested.

Organization Size		Technical Innovations[a]	Administrative Innovations[a]
Small	(N = 37)	25.54	19.58
Medium	(N = 38)	34.78	26.12
Large	(N = 10)	53.44	41.43
Total	(N = 85)	32.95	25.08

Source: Damanpour, Evan (1984, p. 402).

[a]   The rate of adoption of innovations is measured by the percentage of innovations adopted from the total innovations (of each type) available for adoption.

Table 1: Comparison of the rate of adoption of technical and administrative innovations in library organizations

2.1.2 The dual-core model

The dual-core model proposes that the adoption of administrative and technical innovations follow two distinct processes. According to this model, organizations have a technical core and an administrative core. The technical core is primarily concerned with the transformation of raw materials into organizational products and services, while the administrative core's main responsibilities are the organizational structure, control systems, and coordination mechanisms (Daft 1978). Innovation can occur in each core, but technical and administrative innovations follow different processes. Technical innovations typically originate in the technical core and follow a bottom-up process, while administrative innovations originate in the administrative core and follow a top-down process (Daft 1978).

The dual-core model also suggests that the structures that facilitate innovation in each core are different. A mechanistic structure better allows an organization to adapt to changes in goals, policies, strategies, structure, control systems, and personnel (Daft 1982). Thus, low employee professionalism, high centralization in decision making, and high formalization of behavior facilitate the top-down process of administrative innovations. On the other hand, an organic structure facilitates changes in organizational products, services, and technology (Daft 1982). Thus, high professionalism, low centralization, and low formalization facilitate the bottom-up process of technical innovation. The dual-core model, therefore, proposes that the appropriate organizational structure for innovation could be either mechanistic or organic (Burns, Stalker 1961) depending on the type of innovation that needs to be adopted.

2.1.3 Dominant innovation issue

The relative balance between the adoption of administrative and technical innovations depends upon environmental conditions, organizational goals, and whether the "dominant innovation issue" faced by the organization pertains to the administrative or technical domain of the organization (Daft 1982). According to Daft, the external environment can be conceived of as having two parts: (1) the "administrative sub-environment", which pertains to the administrative component of the organization, and includes the community context, resource granting agencies, political and social factors, and government organizations; and (2) the "technical sub-environment", which pertains to the technical system of the organization, and is composed of competitors, customers, suppliers, and technical groups (1982, pp. 150-151). When the administrative sub-environment is complex and dynamic, the dominant innovation issues are administrative and the rate of administrative innovation is high. On the other hand, when the technical sub-environment is complex and dynamic, the dominant innovation issues are technical and the rate of technical innovations is high.

In an indirect test of Daft's model, Damanpour, Szabat, and Evan (1989) compared performance of organizations that had a low rate of adoption of technical innovations and a high rate of adoption of administrative innovations (dominant innovation: administrative) with those that had a high rate of adoption of technical innovations and a low rate of adoption of administrative innovations (dominant innovation: technical) in two five-year time periods. While these researchers found a change in the mean performance measures of organizations in the two periods consistent with the theory, the difference between the mean performance measures was not significant at the 0.05 level in either of the two periods.

2.2 Product and process innovations

Product innovation is defined as new products or services introduced to meet an external user or market need, and *process innovation* is defined as new elements introduced into an organization's production or service operations (e.g., input materials, task specifications, work and information flow mechanisms, and equipment) to produce a product or render a service (Ettlie, Reza 1992; Knight 1967; Utterback, Abernathy 1975). Product innovations have a market focus and are primarily customer driven, while process innovations have an internal focus and are mainly efficiency driven (Utterback, Abernathy 1975). The distinction between product and process innovations is important because their adoption requires different organizational skills: product innovations require that firms assimilate customer need patterns, design, and manufacture the product; process innovations require firms to apply technology to improve the efficiency of product development and commercialization (Ettlie et al. 1984). Patterns of adoption of product and process innovation have more widely been studied at the industry than the firm level of analysis. We review the studies at both levels.

2.2.1 Product cycle model

The product cycle model, perhaps the most widely cited model of product and process innovations, was developed by Abernathy and Utterback (1978). The model, conceived at the industry level, describes the changing rates of product and process innovations over three phases of the development of a product class. In the first phase, the "fluid phase", the rate of product innovations is greater than the rate of process innovations. The flurry of product innovations eventually end with the emergence of a dominant design (Utterback 1994). During the second phase, the "transitional phase", the rate of product innovations decreases and the rate of process innovations becomes greater than the rate of product innovations. In this phase, while the emergence of the dominant design reduces product variety, efforts shift toward producing the standard product more efficiently (Utterback 1994). Finally, in the third phase, the "specific phase", the rates of both types of innovations slow down and become more balanced (Abernathy, Utterback 1978). The first two phases are periods of radical change, where major product innovations and major process innovations are introduced respectively; the third phase is a period of incremental change, where less fundamental product and process innovations are introduced at more congruent rates (Abernathy, Utterback 1978). The product cycle model was empirically supported by the data from the Myers and Marquis's (1969) study of successful technological innovations in five industries (Utterback, Abernathy 1975) and case studies of the development of several product classes (Utterback 1994).

The Abernathy-Utterback model focuses on a single cycle of technological change. More recent studies of the history of industries suggest that technological change is cyclical; that is, "dematurity" can return an industry from the specific phase to the fluid phase (Anderson, Tushman 1991). A "discontinuous change" (Tushman, Anderson 1986) or an "environmental jolt" (Meyer 1982) can lead to a new series of product and process innovations in an industry. For example, the U.S. banking industry experienced an environmental jolt when several major laws were introduced between 1978 and 1982. These laws deregulated the industry, created more competition, and increased the introduction of both product and process innovations (Damanpour, Gopalakrishnan 1998b).

2.2.2 Reverse product cycle model

Barras (1986, 1990) developed a model of product and process innovations for service or user (technology adopting) industries. He argues that the product cycle model applies to the production of goods embodying a new technology in the goods (technology supply) industries. In the user industries, which usually adopt the technology developed in goods industries, the cycle, which he terms the "reverse product cycle", operates in the opposite direction. In the first phase, the technology is used to increase the efficiency of existing services; in the second phase, it is applied to improve the quality and effectiveness of services; and in the third phase, it assists in generating wholly transformed or new services (Barras 1986). According to this model, innovations emphasized by the adopting organization in the service industries are, respectively: incremental process innovations to increase efficiency; radical process innovations to improve effectiveness; and radical product innovations to generate new services (Barras 1986).

Barras (1990, pp. 227-231) illustrates the reverse product cycle using the introduction of information technology in retail banking from 1960s to 1990s. During mid 1960s - mid 1970s (first phase), the mainframe computers became commercially available and banks used this technology to automate transactions and financial records to incrementally increase the efficiency of back-office operations. During mid 1970s - mid 1980s (second phase), when new developments in computer technology offered electronic funds transfer, banks invested in corporate networks linking dumb terminals in the form of Automated Teller Machines (ATMs) to their central computers. This technology enabled banks to automate a part of the direct service delivery function, thus improving the quality of services offered to their customers. From mid 1980s onward (third phase), further developments in computer technology allowed banks to apply interactive and integrated computer networks with intelligent terminals. When interactive networking is established between banks, shops, and homes, banks can offer new services to their customers, such as personal financial services (including mortgage lending, insurance, taxation, and investment advice), home banking, and cashless shopping.

2.2.3 Pattern of adoption at firm level

Barras's model suggests a different pattern of adoption of product and process innovations than Abernathy and Utterback's model. However, both models were developed at the industry level of analysis. At the firm level, while product and process innovations have been studied separately (Dougherty, Hardy 1996; Ettlie, Reza 1992; Hambrick, MacMillan, Barbosa 1983; Schroeder 1990), studies of the relative rates of adoption of product and process innovations are rare. An exception is a study of the adoption of product and process innovation in the banking industry (Damanpour, Gopalakrishnan 1998b). We studied 31 innovations adopted by a sample of 101 U.S. commercial banks between 1982-1993 and found that banks, on average, adopt product innovations at a higher rate and with a greater speed (i.e., earlier) than process innovations. This was true for banks of all size categories - small, medium and large (Table 2).

Organization Size	Rate of Adoption[a]		Speed of Adoption[b]	
	Product	Process	Product	Process
Small (N = 42)	41.04	20.92	2.55	2.16
Medium (N = 45)	53.46	27.46	3.00	2.40
Large (N = 14)	79.41	53.57	3.32	2.58
Total (N = 101)	51.89	28.36	2.86	2.33

Source: Damanpour, Gopalakrishnan (1998b).

[a] Rate of adoption equals the percentage of innovations adopted by a bank from the total innovations available for adoption.

[b] High speed score represents early adoption of innovations.

Table 2: Comparison of the rates and speeds of adoption of product and process innovations in commercial banks

2.3 Radical and incremental innovations

Like the other typologies of innovation, the proposed theories of innovation radicalness refine the term innovation by dividing it into two separate kinds. Several variations on these paired kinds for innovation radicalness are proposed by researchers. For example, Normann (1971) used the terms "variation" and "reorientation"; the former implies refinements and modifications to existing products, whereas the latter implies fundamental changes. Knight (1967) and Nord and Tucker (1987) distinguished between "routine" and "non-routine" innovations depending upon whether the innovation produces minor or major changes in products, services, or production processes in the organization. Grossman (1970) distinguished between "ultimate" innovations - those that are ends in themselves - and "instrumental" innovations - those that facilitate the adoption of ultimate innovations at a later point in time. In empirical research, these categories are often collapsed into the terms radical and incremental innovations, which we use here. *Radical innovations* produce fundamental changes in the activities of the organization and represent clear departure from existing practices, whereas *incremental innovations* result in a lesser degree of departure (Dewar, Dutton 1986; Ettlie et al. 1984).

As in the case of product and process innovations, radical and incremental innovations have been studied at both the industry and the firm level of analysis. While industry level studies primarily emphasize the development of these innovations types, firm level studies mainly emphasize their adoption. The majority of studies in both groups distinguish between the parameters that influence the introduction of radical versus incremental innovations, and the studies are primarily conceptual.

2.3.1 Industry level parameters

Three types of industry level parameters differentially impact the introduction of radical and incremental innovations - industry life cycle, industry networks, and the distinction between incumbent and newcomer firms.

For industry life cycle, the findings of the Utterback and Abernathy's (1975) study can be reinterpreted for radical and incremental innovations. As stated earlier, in the "fluid phase" of an industry's life cycle, there are competing product technologies and the emergence of a radical technology, or a "dominant design", marks the end of this phase (Anderson, Tushman 1990). The "transitional phase" is marked by the emergence of radical process technologies which lower costs and improve efficiencies. The "specific phase" is characterized by introduction of incremental product and process innovations which lead the industry through a period of relative stability and consolidation. Thus, there is a greater emphasis on radical technologies in the emergence and growth phases, and incremental technologies in the maturity phase, of the development of an industry or product class.

Industry networks and collaborative alliances also differentially influence the introduction of radical and incremental innovations. For radical innovations, Rycroft and Kash (1994) argued that the development of complex technologies requires "interconnectedness", which is provided by networks of organizations. Organizational networks facilitate the development of radical or complex technologies because different components or subsystems of innovation are developed by different firms (Tidd 1995). Lee (1995) substantiated this claim when he found that technical linkage with buyers and suppliers had a stronger effect on radical innovation in a new technology setting than in a traditional setting. In addition to external linkages, organizational networks influence internal lines of communications. The players in the network influence the process of consensus building, the selection of courses of action, and above all, they behave as a community, rather than a random aggregation of actors. For the introduction of incremental innovations, however, the importance of networks is significantly less than that of radical innovations. Firms seem to be able to independently develop and implement incremental innovations. These innovations tend to be more autonomous (Chesborough, Teece 1996), more "organization specific" and less complex (Damanpour 1996), and therefore, amenable to simpler development processes when compared to radical innovations that are more systemic and complex.

Finally, it has been found that technology leaders or large incumbents within the industry favor exploiting existing technologies through incremental innovations (Kusunoki 1997), whereas the smaller players or the industry outsiders often develop radical new technologies (Bower, Keogh 1996). For example, in the early 1970s, Matsushita had been holding a leadership position in a variety of product technologies including digital data processing over competitors such as Ricoh, Hitachi, and NEC (Kusunoki 1997). However, Matsushita lagged behind Ricoh, which succeeded in developing the first digital facsimile machine in 1973 (Kusunoki 1997). Tushman and Anderson (1986) in a study of innovations in airline, cement and minicomputer industries also found that competence destroying or radical technologies are initiated by new firms and are associated with increased environmental turbulence, while competence-enhancing technologies are initiated by existing firms and are associated with decreased environmental turbulence.

To summarize, at the industry level, radical innovations are more likely to occur in the earlier stages of an industry's life cycle, are facilitated by open networks among organizations, and are developed by relative newcomers to the industry. Incremental innovations are more likely to occur in the maturity phase of industry life cycle, are relatively stand-alone, and are developed by large incumbents to strengthen their existing position in products or process technologies.

2.3.2 Firm level parameters

Empirical research that distinguishes between predictors of radical and incremental innovations at the firm level is scarce, and a dominant theory to distinguish between these two types of innovation has not yet emerged. Studies that examine the predictors of radical and incremental innovations have analyzed either structural or market-related factors that may impact their adoption.

Structural factors that have been examined include centralization, specialization, integration, and the physical location of innovation projects within the organization. Hage (1980) argued that innovative organizations with organic structures would innovate incrementally because they have more democratic values and power is shared, whereas organizations with mechanistic structures may be a fertile ground for radical change (Nord, Tucker 1987). Dewar and Dutton (1986) hypothesized that favorable managerial attitude toward change, concentration of technical specialists, the depth of the organization's knowledge resources facilitate radical innovations but have no association with incremental innovations. On the other hand, they argued that centralization in decision making positively affects radical and negatively affects incremetnal innovations (Dewar, Dutton 1986). Germain (1996) found that while decentralization of decision making favored incremental innovation, specialization enhanced the adoption of both radical and incremental innovations. Ettlie and colleagues (1984) found that radical innovations are more likely to occur in organizations with centralized and informal structures, while incremental innovations are more likely to occur in organizations with complex and decentralized structures. In sum, the results are mixed and do not consistently support the advanced theories.

In terms of location of innovation projects, it has been found that radical innovation projects need to be sheltered from the pressures that face normal business (Port, Carey 1997) and be supported with an intensive investment process (Lynn, Morone, Paulson 1996). In a study which analyzed major technological breakthroughs, Lynn and colleagues (1996) found that GE in its commercialization of the CT scan machines, Motorola in its cellular phone program, and Corning in its invention and marketing of the optical fibers faced numerous adverse market reactions and technological problems which could only be overcome by sustained resource investments and managerial persistence. However, contrary to the development process of radical innovations, incremental innovation projects were much more integrated to the existing activity of firms and had a shorter lag time before the firms experienced benefits from them (Lynn et al. 1996).

Considering the market-related factors, researchers have argued that the acceptability of each of these types of innovations to the customers, and hence their subsequent success, is affected by different factors. For example, while to market radical innovations successfully organizations should primarily focus their marketing efforts on relative advantage and observability of the new product or service, to market incremental innovations, they should focus on the economic and performance advantages and the ease of use associated with the new product or service (Strutton, Lumpkin, Vitell 1994). Thus, different innovation attributes may need to be emphasized in the marketing and commercialization processes based on whether an innovation is perceived as radical or incremental by the customer.

2.4 Empirical results of differences between innovation types

During 1980s and 1990s, the three typologies of innovation presented above were empirically examined. Individually, many of these studies reported some differences between innovation types (Daft 1978; Damanpour 1987; Kimberly, Evanisko 1981). However, Damanpour's (1991, 1992) meta-analytic reviews of the determinants of innovation at the organizational level of analysis, provided an opportunity to examine the differences between types of innovations across these studies. Table 3 reports mean correlations between ten structural and process variables and the three typologies of innovations. With the exception of the effect of specialization on administrative versus its effect on technical innovations, there were no other statistically significant differences at the 0.05 level between the mean correlations of any of the three types (Table 3). That is, organizational variables that predict the adoption of one type of innovation also predict the adoption of the other type. The data in Table 3, therefore, suggest that the aggregated results of past empirical studies do not support the intermediate theories of organizational innovation depended on differences between innovation types.

Predictor Variable	Innovation Type					
	Adminis-trative	Technical	Product	Process	Radical	Incre-mental
Size	.27	.29	.57	.46	.54	.45
Specialization[a]	.21	.53	.58	.55	.56	.53
Functional differentiation	.29	.38				
Professionalism	.17	.13	-.02	.30	.31	.34
Fomalization	-.07	.04	.29	.16	-.10	-.04
Centralization	-.11	-.19	-.21	-.06	-.12	-.13
Vertical differentiation	.28	.03				
Managerial attitude toward change					.10	.18
Technical knowl-edge resources			.36	.52	.59	.48
External communication	.38	.36			.27	.37

[a] t-approximation test of difference between mean correlations of administrative and technical innovation is significant at the 0.05 level. Source: Damanpour (1991, p. 577).

[b] Entries are mean correlations between predictor variables and organizational innovation from published empirical studies. Source: Damanpour (1991, pp. 570-573); and Damanpour (1992, p. 384).

Table 3: Research results for the organizational determinants of innovation types[b]

In this paper, as stated earlier, we take a different view of useful-ness of innovation types. Typologies of innovation are necessary because types of innovation relate differently to different subsystems of the organization. Further, systems theory suggests that: (1) a change in one subsystem of an organizational system requires ad-justments in the other related subsystems, and (2) the overall per-formance of the organizational system depends on how well the sub-systems fit together and work with each other (Ackoff 1981; Emery, Trist 1960; Galbraith, Nathanson 1978). This view is compatible with our conception of innovation as a means of organizational change and adaptation. Based on this view, in the next section, we argue that congruency in the adoption of innovation types in organi-zations contributes to organizational adaptation and effectiveness.

3. Congruency in the adoption of innovation types

3.1 Adoption of innovation

Organizations can generate and/or adopt innovations (Gopalakrishnan & Damanpour, 1994). Innovations that are *generated* in an organization are either for internal use or for sale to other organizations. The generation process results in an outcome - a new product, service, program, or technology. If this outcome is then acquired by another organization, or by a different part of the generating organization, the second organization goes through another process, that of *adopting* the innovation. For the generating organization, the innovation process entails idea generation, project definition, design and development of the product or service, and marketing and commercialization (Rothwell, Robertson 1973; Baker, McTavish 1976; Cooper, Kleinschmidt 1990). For the adopting organization, the innovation process includes awareness of innovation, attitude formation, evaluation, decision to adopt, trial implementation, and sustained implementation (Rogers 1995; Zaltman et al. 1973). In this paper, we focus only on the adoption of innovation.

The adoption of innovation is presented by either the rate or the speed of adoption (Damanpour, Gopalakrishnan 1998a). The *rate of adoption* relates to the extent of innovativeness of the organization. It reflects the magnitude of innovation, i.e., the number of innovations the organization adopts within a given period. Organizations with a high adoption rate adopt innovations more frequently and more consistently. Innovation researchers have typically assumed that both the extent of organizational innovativeness and pervasiveness of innovation across organizational units would help organizational effectiveness and competitiveness. The *speed of adoption* relates to the timing of innovation, i.e., the speed with which the organization adopts innovations after their first introduction elsewhere. It reflects the organization's responsiveness and its ability to adopt innovation quickly relative to its competitors within the industry (Lengnick-Hall 1992). The speed of adoption differentiates between early and delayed adoptions, which, in turn, has consequences for organizational performance. Researchers have often associated early adoption of innovation to strong organizational performance (Kessler, Chakrabarti 1996; Lawless, Anderson 1996; Lieberman, Montgomery 1988). In this paper, we use rate of adoption, the dimension more commonly used in empirical studies of organizational innovation.

3.2 Innovation adoption and organizational adaptation

As stated earlier, the performance gap concept provides a conceptual underpinning for the adoption of innovation as a means of organizational adaptation. The *performance gap* is the discrepancy between what the organization is actually doing and what its executives believe it can potentially do (Zaltman et al. 1973). It is the difference between the executives' criteria of satisfaction and the actual performance of the organization. The perceived performance gap creates a need for change in the organization. Change is more likely when performance is below the aspiration level of the top executives (Mezias, Lynn 1993). The need for change stimulates the adoption of innovations which in turn facilitates the organization to close the gap (Zaltman et al. 1973). Consequently, the performance gap concept posits that: (1) innovative organizations respond to perceived performance gap by adopting innovations; and (2) the adoption of innovation closes the gap and contributes to the effectiveness of the organization.

From a system perspective, performance is the ability of an organization to cope with all four systemic processes (inputs, outputs, transformations, and feedback effects) relative to its goal-seeking behavior (Evan 1976). A high-performance organization would accomplish its primary tasks efficiently and would carry out its organization-maintaining and organization-adapting functions effectively (Miles 1980). The organization adapting function requires that, as the external or internal environment change, the structure and processes of the organization undergo change to meet the new environmental conditions. Some organizations tend to do more, however. They not only adapt to environmental conditions, but also use their resources and skills to create new environmental conditions by introducing new products, technologies, or services never offered previously.

Whether internally or externally induced, reactive or proactive, the adoption of innovation has consequences for the adopting organization. *Consequences of innovation* are the changes that occur to an organization as a result of the adoption of innovations (Rogers

1995). Consequences are difficult to measure in a precise manner because they are not always "direct" or "anticipated", but may be "indirect" or "unanticipated" (Goss 1979; Rogers 1995). A direct change introduced in one part of the organization as a result of the adoption of an innovation may produce indirect and unanticipated changes in that or the other parts of the organization. The impact of innovation adoption on organizational performance would be a result of both direct and indirect, and anticipated and unanticipated consequences. The consequences of innovation would be desirable if the organization as a whole can cope with and benefit from the changes in all its parts. As stated earlier, different types of innovation influence different parts or subsystems of the organization; therefore, the congruency in the rate of adoption of innovation types influences organizational adaptation.

3.3 Dynamics of innovation types and organizational adaptation

How would the type of innovation affect the relationship between the rate of adoption of innovations and organizational performance? Researchers have argued that perceived effectiveness of innovation types are not the same; for example: (1) Damanpour and Evan (1984) argued that technical innovations are perceived to be more effective than administrative innovations; (2) Frost and Egri (1991) proposed that product innovations are perceived more effective than social innovations; (3) Damanpour and Gopalakrishnan (1998b) argued that product innovations are perceived more effective than process innovations; and (4) Gopalakrishnan and Bierly (1997) suggested that complex or radical innovations are perceived to be more effective than simple or incremental innovations. However, innovations that are perceived more effective may not objectively influence organizational performance, because the perceived effectiveness depends not only on the innovation's potential contribution to performance, but also on the social status, visibility, the immediacy of rewards, and compatibility of the innovation (Rogers 1995). The arguments for perceived effectiveness of one type of innovation over another notwithstanding, we focus on the synchronous adoption of innovation types as an important factor for organizational adaptation.

3.3.1 Dynamics of administrative and technical innovations

Damanpour and Evan (1984) employed the sociotechnical systems framework to argue for the congruency in the rate of adoption of administrative and technical innovations. According to these authors, technical innovations are not merely innovations resulting from the use of technology. They are innovations that occur in the technical system of an organization and are directly related to primary work activity of the organization; hence, they are a means of changing and improving the performance of the technical system of the organization. Administrative innovations are innovations in organizational structure and the management of people (Knight 1976). They occur in the social system of an organization. The social system here refers to the relationships among people who interact to accomplish a task or a goal. It includes those rules, roles, procedures, and structures that are related to the communication and exchange among people and between people and their environment (Cummings, Srivastva 1977). Administrative innovations are a means of changing and improving the performance of the social system of an organization.

The sociotechnical system theory emphasizes the role of both technical and social systems operating jointly for the effective operation of an organization. Sociotechnical theorists argue that concentration on either the technical or the social system without due regard to the other would result in low organizational performance and growth (Herbst 1974). The social and technical systems of an organization need to function in balance if the organization as a whole is to operate effectively. The relationship between the social and the technical system is not strictly a one-to-one relationship, rather it is a correlative one (Emery, Trist 1960). For example, any change in the technical system sets certain constraints and requirements for the social system. The effectiveness of the total organizational system depends on how adequately the social system copes with these requirements. If the social system is not prepared and cannot adjust to the demands of the technical system, the required match between the two systems for high performance of the organization would not be achieved.

Because organizational performance is a direct function of the balance between the social and technical systems, a discrepancy in the rate of adoption of administrative and technical innovations would have an adverse effect on organizational performance (Damanpour, Evan 1984). In an empirical study of library organizations, Damanpour and Evan (1984) found that the association between technical and administrative innovations is stronger in high-performance than in low-performance libraries. Ettlie (1988) found similar results in the manufacturing sector. He reported that: (1) successful firms adopt administrative and technical innovations simultaneously; and (2) the congruency between the two types of innovation is especially important during hostile and competitive times.

3.3.2 Dynamics of product and process innovations

We proposed and tested two views of the pattern of adoption of product and process innovations: (1) "the lag pattern", which posits that, over time, organizations adopt product innovations followed by process innovations to enhance the previously adopted product innovations; and (2) "the synchronous pattern", which proposes that organizations adopt product and process innovation simultaneously (Damanpour, Gopalakrishnan 1998b). Both product cycle model and reverse product cycle model assume a lag pattern of introduction of product and process innovations (Abernathy, Utterback 1978; Barras 1986); that is, the introduction of one type of innovation lags or leads the introduction of the other type over time. At the industry level, a lag pattern may suit the early stages of the development of an industry or a product class. At the firm level, however, we posit that a synchronous pattern of adoption of product and process innovations enhances organizational effectiveness more than a lag pattern.

In competitive environments, innovation adoption is stimulated by simultaneous pressures to increase efficiency (reduce costs) and effectiveness (improve quality). Therefore, effective organizations tend to adopt both product and process innovations to maintain high performance. Ettlie (1995) found that integrated product-process development practices have positive implications for firm performance. Pisano and Wheelwright (1995), in a study of U.S. and European pharmaceutical companies, argued that the simultaneous development of new products and new processes is not only possible but necessary. They reported that companies in a variety of high technology industries had gained tremendous advantages by treating process development as an integral part of product development.

The congruent adoption of product and process innovations resulted in a smoother launch of new products, easier commercialization of complex products, and more rapid penetration of markets (Pisano, Wheelwright, 1995). In service industries, more than the manufacturing ones, it is difficult to separate new products from the processes on which they rely on; thus, the introduction of new products require the introduction of new production processes simultaneously (Buzzacchi, Colombo, Mariotti 1995). In a study of innovations in the banking industry, we found a synchronous pattern of adoption of product and process innovations is more effective than a lag or lead pattern (Damanpour, Gopalakrishnan 1998b). Specifically, we compared low-performance (bottom quartile) and high-performance (top quartile) banks in our sample using both ROA and ROE, and found that the association between product and process innovations was statistically significant ($p < 0.05$) in high-performance but not in low-performance banks.

3.3.3 Dynamics of radical and incremental innovations

Unlike in the case of administrative and technical and product and process innovations, no empirical study has compared the effectiveness of the lag versus the synchronous pattern in the adoption of radical and incremental innovations. However, Tushman and O'Reilly (1997) argued that the route to sustained competitive advantage is not through succeeding at either radical, architectural, or incremental innovations, but through adoption of streams of innovations as the external environment changes and the market unfolds. *Streams of innovations* constitute multiple innovations, of different degrees of radicalness, through which a firm simultaneously reaps the benefits of periods of incremental change and shapes the pace and direction of radical innovations needed for periods of discontinuous change (Tushman, O'Reilly 1997, p. 166).

The correlations between radical and incremental innovations in empirical studies at the firm level also suggest that organizations that have a propensity to adopt radical innovations also adopt more incremental innovations (Dewar, Dutton 1986; Ettlie et al. 1984; Germain 1996). Organizations that have a structure and culture that is favorable for innovation seem to be more adaptable and capable of adopting different types of innovation. This is further corroborated by evidence from a case study of three multi-product diversified companies - Asea Brown Boveri (ABB), Hewlett-Packard (HP), and

Johnson & Johnson (J&J) - which shows the ability to pursue both incremental and radical innovation leads to long-term success (Tushman, O'Reilly 1996). The three companies have been able to compete through incremental innovations in mature markets and through radical innovations in emerging markets. Therefore, to sustain high performance and succeed in the marketplace, firms need to be competent at both radical and incremental innovations. ABB, HP, and J&J have devised structure and processes that enable them to adopt these innovation types synchronously.

3.4 Empirical results of associations between innovation types

At the organizational level of analysis, with few exceptions stated earlier, the dynamics of the adoption of innovation types has not been examined empirically. However, a review of the studies that have investigated more than one type of innovation and have reported the correlations between innovation types demonstrates that a positive and significant association between the rate of adoption of pairs of innovation types is common (Table 4). This is perhaps the most consistent finding in innovation research!

The results presented in table 4 generally support our contention that innovative organizations adopt pairs of innovation types congruently. The value and impact of innovation for organizational effectiveness become apparent when organizations adopt innovations in related sets, where each innovation's contribution is complemented and enhanced by the adoption of other innovations (Rosenberg 1982). The pairs of innovation types discussed in this paper are usually linked in such a way that the adoption of one type requires the adoption of the other type for best impact. Organizational performance is a function of innovating, not adopting radical, technical, product or any one type of innovation alone. The management of innovation, therefore, entails managing streams or sets of innovation, which requires a new perspective in both studying and managing innovations in organizations. Innovation should be considered as a managerial process rather than a purely technological process (Mezias, Glynn 1993). To successfully adopt sets of innovations, managers should be able to understand and cope with the different requirements of each type of innovation and introduce structures, cultures, and systems that facilitate the synchronous adoption of multiple types (Tushman, O'Reilly 1997).

Study	Administrative and technical	Product and process	Radical and incremental
Bantel & Jackson (1989)	.55*		
Damanpour (1987)	.39***		
Damanpour & Evan (1984)	.42** .37** .47**		
Damanpour & Gopalakrishnan (1998b)		.59*** .47*** .43***	
Dewar & Dutton (1986)			.61***
Ettlie, Bridges, O'Keefe (1984)		.31** .43**	.54**
Germain (1996)			.35**
Kimberly & Evanisko (1981)	.42***		
Zahra & Covin (1994)	.29** .21*	.31***	

[a] Entries are zero-order correlations between pairs of innovation types.

* p≤.05; ** p≤.01; *** p≤.001.

Table 4: Associations between pairs of innovation types[a]

4. Concluding remarks

We examined the differences between three categories of innovation - technical versus administrative, product versus process, and radical versus incremental - and the patterns of their adoption at the organizational level. Although two competing models of innovation adoption have been proposed for these types of innovation - the lag model, which favors sequential adoption, and the synchronous model, which favors concurrent adoption - we suggest that the synchronous model generally has more favorable consequences for organizational adaptation and effectiveness. In other words, organizations that adopt technical and administrative, product and process, or radical and incremental innovations congruently generally are able to achieve better internal alignment among their various systems, structure and processes, and this, in turn, influences their performance positively. We examine the implications of these results for both future research and practice of innovation management.

With respect to future research, because few empirical studies have tested the synchronous model (Damanpour, Evan 1984; Damanpour, Gopalakrishnan 1998b; Ettlie 1988), we recommend more research at the organizational level to establish the existence and consequences of congruency in the adoption of innovation types. The effectiveness of synchronous adoption may be dependent on contextual variables like size, environmental uncertainty, and technological complexity. These contextual factors can influence the adoption of sets or streams of innovations, which future research can tease out. For example, larger organizations may benefit more from synchronous adoption of innovation types than smaller organizations because they have a greater need for administrative innovations to structurally coordinate activities, process innovations to improve efficiency, and incremental innovations to generally achieve internal alignment and improve their existing methods of doing business. While larger organizations have greater financial and human resources to devote to simultaneous adoption of innovation sets, smaller organizations, because of limitation of resources, and simplicity of their structure, may benefit more by focusing on one type of innovation within each category.

Further, in this paper, we have analyzed patterns of adoption only within a category - technical versus administrative, or product versus process. Future research can extend our work by examining the patterns of adoption across categories. For example, do organizations adopt radical administrative innovations synchronously with incremental technical or radical technical innovations? What factors would influence the relationship between the adoption of different sets of innovation and organizational performance? Mezais and Glynn (1993) suggested three approaches to study innovations - institutional, revolutional, and evolutional. Strategies for managing sets of innovation in each of these approaches are fundamentally different. Future research can identify the innovation set-organizational performance linkage within each approach.

From a practitioner's point of view, the importance of congruency of adoption of innovation types has implications for designing the structure and processes, as well as the evolution of organizational leadership and culture. For example, technical and radical innovations are facilitated by organic structures and administrative and incremental innovations by mechanistic structures (Daft 1992; Tushman, O'Reilly 1997). To achieve congruency in the adoption of different types of innovation, therefore, organizations have to develop and manage ambidextrous structures. One way to do this would be through matrix structure and horizontal linkage mechanisms like task forces or project teams (Daft 1992). Organizational members, in addition to occupying the position afforded by the structure would also frequently be a part of task forces, project teams or implementation committees. Such a structural mechanism would provide the benefits of mechanistic aspects of structure such as spatial, hierarchical, occupational, and functional differentiations (Miller, Contay 1980), and through the horizontal linkages, would also provide the features of a small, autonomous project group or unit that is more organic in nature.

The top management is largely responsible for the cultural values that prevail in support of innovation within an organization (Bantel, Jackson 1989). The culture in organizations where congruency of the adoption of innovation types is encouraged should be simultaneously tight and loose. It should be tight in the sense that there are broadly shared norms that are critical for innovation such as openness, autonomy and risk-taking, and it should be loose in the manner in which these common values expressed varies on the type of innovation required (Tushman, O'Reilly 1996). The culture for innovation should promote identification and sharing of information and resources and also provide consistency that promotes trust and predictability. A culture that has the qualities of tightness and looseness promotes a general climate for innovation by enabling the structure to be ambidextrous. A flexible culture and an ambidextrous structure together enable the organization adopt a congruent pattern of innovation types to adapt to the diverse environmental contingencies, which is conducive for sustained, high organizational performance.

Strategic dynamics and corporate performance: A longitudinal assessment

Jens Leker

Strategic dynamics and corporate performance: A longitudinal assessment

Jens Leker

1. Introduction

The analysis of the relationship between strategic change and corporate performance has already been a topic of research. An up-to-date discussion of the respective results can be found in Parnell (1998, pp. 20-24). Looking at the research results more closely, we notice that performance - which is usually measured in terms of RoA, RoI, or the development of turnover - is modeled both as a dependent as well as an independent variable, depending on the aim of the research. One question addressed is how the undertaking of a strategic change can be explained by the development of a firm's economic situation. In such a case, corporate performance is considered to be an independent variable. Another issue being analyzed is how the economic situation of a business is influenced by a strategic change. In this case, corporate performance is considered to be a dependent variable.

Summarizing the results of these studies into a chart, we are able to determine the following:

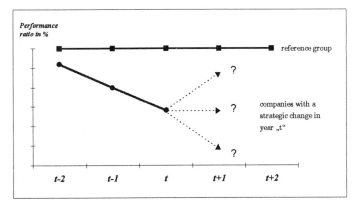

Figure 1: Strategic dynamics and corporate performance

The majority of researchers who examine the influence of the economic situation on strategic change (Boeker 1989, p. 507, Webb, Dawson 1991, p. 203; Parnell 1998, p. 30; Greve 1998, pp. 80-82) conclude that the likelihood of a strategic change occurring at companies showing unsatisfactory corporate performance is significantly higher. Moreover, this likelihood increases with a deterioration of the economic situation. This finding is supported by theories of risk taking and aspiration levels (Greve 1998, pp. 59-66)

However, the studies analyzing the impact of strategic change on corporate performance do not, as a matter of fact, come to clear positive findings. One is, thus, confronted with research results claiming that a strategic change leads to an improvement in corporate performance. At the same time there are studies observing a further deterioration or comparatively inferior development in the strategically altered companies compared to businesses maintaining a uniform strategic direction (Gannon, Smith, Grimm 1992, pp. 237-239; Parnell 1994, p. 25; Parnell 1998, p. 30).

Seen in relation to this research background, the question seems justified whether or not most of these contradictions could be attributed to lack of differentiation. This is the point at which my research picks up the investigation. It is examined whether it is correct to speak of "*the*" strategic change - in the sense of a 0/1-variable - or whether it might not be possible to discover different types of strategic changes, accompanied by different manners of corporate performance development.

2. Differentiation of strategic dynamics - a typology of strategic change

2.1 Data

To analyze the research question, it is necessary to define strategic change. In the framework of my research, we refer to a "strategic" change if this change has been specifically announced by the company's management and if it has, at the same time, been credibly announced to the firm's bank. This precondition leads to the fact that these strategic changes analyzed can, for the most part, be characterized as "radical" strategic changes, if we evaluate the extent of the changes on a continuous scale ranging from minor or incremental

(Lindblom 1959; Quinn 1980) to radical or fundamental strategic changes (Miller, Friesen 1980; Tushmann, Romanelli 1985).

The above definition corresponds with the research relying on interviews with managers or on questionnaires addressed to top management (Snow, Hambrick 1980, pp. 532-536). Here, however, the strategic change is identified on the basis of document analysis - an analysis of the credit file of the firm in question, to be exact. For this reason, self-assessment by the top management is in no way related to or influenced by the aim of our study. The credit files made available for our research contained mainly documents of general credit correspondence, balance sheets, records of solvency analysis, reports, registers and records of securities. The detailed credit reports, in particular, served as a source for identifying the framework of a new strategic direction. In this case, the plausibility test for additional confirmation of the top management's self-assessment (which would normally be essential under other research circumstances) is replaced by the bank's evaluation found within the credit reports.

Beginning with the customers of a major German bank which co-operated in this study, an analysis of all credit committee protocols for the period from November 1, 1993 through November 20, 1995 was possible - this two year time span represents the maximum possible prolongation period. Nearly 180 companies could be identified whose protocols included indications of a new strategic direction. Analysis of the credit files led to a sample of 65 major corporations which, according to their own statements and confirmed by the bank, undertook a strategic change. Table 1 shows the legal structures of these companies and their industries.

Legal forms:	n
Public company (AG) (on stock exchange)	30 (24)
Company with limited liability (GmbH)	18
Mixed associations (i.e. GmbH & Co.KG)	12
Limited partnership (KG)	4
Partnership (OHG)	1
Industries:	n
Mechanical engineering	11
Electrotechnology	7
Chemical industry	6
Trade / Merchants	6
Technology of measuring and controlling	5
Paper industry	4
Automobile / Automobile contractors	4
Other	22

Table 1: Sample description

2.2 Method

To characterize the individual strategic changes, the credit files of the 65 corporations were analyzed in detail under the following criteria (Hauschildt 1997, pp. 7-16):

1.) the motives cited for the strategic change,

2.) the objectives and defining activities around the strategic change,

3.) the bank's evaluation in terms of the risks of such a strategic change.

Table 2 presents the variables for the each of the above areas. These 18 variables serve as a basis for analyzing the degree to which the identified strategic changes have similar characteristics. A hierarchical cluster analysis for binary variables (Norusis, SPSS Inc., 1994 pp. 106-107; Bacher 1994, pp. 200-209; Backhaus et al. 1990, pp. 118-125) was then performed in order to develop a typology of strategic change.

Motives of the new strategic direction
1. maintaining competitiveness 2. critical situation 3. expansion 4. integration 5. opportunity
Attributes of the strategic change
technology/production procurement: 6. construction 7. reduction 8. shifting to overseas sales sphere: 9. growth 10. reduction 11. globalisation organization: 12. inclusion of external partners 13. change of management 14. change of corporate governance (change of owners) 15. distinctive change of organization 16. distinctive reduction of employees
Risk evaluation of the bank
17. high 18. medium

Table 2: 0/1 variables determining a strategic change typology

2.3 Results

The results of the cluster analysis allow us to group the strategic changes in the 65 corporations we investigated into four different types, as follows:

Type 1 "**Expansionist**" (n=20):
  expansive movement into new industries
  including new external partners

Type 2 "**Innovator**" (n=19):
  movement to new fields of business by building new
  production plants, introducing new technologies and/or
  innovative products

Type 3 "**Reallocator**" (n=11):
  directed towards new and cheaper resources along with a
  reduction of old production facilities

Type 4 "**Concentrator**" (n=15):
  focused on the basic business sectors

Table 3: Typology of strategic change

To determine the number of clusters, we used the values of the fusion coefficients (Norusis, SPSS Inc. 1994, p. 91; Aldenderfer, Blashfield 1984, pp. 53-58). With fairly large increases in the value of the distance measure, these values indicated that either a two-cluster solution (type 1: expansionist/innovator and type 2: reallocator/concentrator) or a four-cluster solution (as described in Table 3) would have been appropriate. Considering the aim of the study we opted for the four-cluster solution, which indicates - with remarkably smaller coefficients - that considerably more homogeneous clusters are being merged than with the two-cluster solution.

These highly simplified descriptions show that the four different types range in their market oriented activities from external expansion to the adoption of new directions internally, from stabilizing changes aimed at new purchase markets to a distinct focus on reduction. All four types are characterized by important differences in how they deal with their core competencies (Prahalad, Hamel 1990; Javidan 1998, pp. 62-70). The expansionist wants to buy and integrate new core competencies, while the innovator is trying to build up and establish new core competencies on his own. The reallocator wants to transfer core competencies to lower wage countries, while the concentrator is trying to focus on the core competencies already established.

A comparison of these empirical types of strategic change with established typologies of strategic types (Miles, Snow 1978; Porter 1980) is restricted. In particular, it is important to note that a typology of strategic types and a typology of strategic change types differ considerably in the subject of investigation and underlying data sets. While the former characterize the existing strategy at a given moment, the latter typify the process change in strategic direction. For this reason, identification of the overall strategies depends on a classification within an existing typology or on an empirical analysis of strategic elements (Brockhoff, Chakrabarti 1988, pp. 170-174; Brockhoff 1990, pp. 455-459; Conant, Mokwa, Varadarajan 1990). The identification of strategic changes relies, instead, on concepts and empirical data that would serve as a basis for evaluating the scope and the direction of the change in strategic course. Consequently, a different typology pattern emerges (Ginsberg 1988). On the other hand, it is promising to compare the intended change-induced strategic position with an existing strategy typology (Damanpour, Chaganti 1990, pp. 231-242; Hambrick 1983).

Drawing primarily on Miles' and Snow's (1978) strategy types, we can roughly characterize the intended strategic position of the expansionist as that of an "Analyzer", or in the words of Damanpour and Chaganti as an "Innovative Diversifier" (1990, p. 237). The innovator, however, might be seen as being closer to the characteristics of a "Prospector".

Similar to a "Defender" in the Miles and Snow typology, the concentrator has chosen to focus on his main business, while the reallocator is also moving into a defender position, but somewhat more actively. Nevertheless, it should be kept in mind that our typology of strategic change is strongly influenced by certain dynamic aspects (Brockhoff 1996, pp. 185-187) that are not delineated in traditional typologies of strategic types.

Making things even more complicated, it ought also to be noted that our credit file analysis lends support to Snow and Hambrick, who stated that: "Managers, however, typically do not think of their organizations as being Defenders, Prospectors, Analyzers, or Reactors. Instead, they may think of their organizations' strategies as resulting from concerns about being biggest, best, first, lowest priced, highest quality, and so forth" (Snow, Hambrick 1980, p. 530). In this respect, we refrain from further and even more subjective interpretations of the strategic change corporations in the light of traditional typologies of strategic types.

The next step is to investigate whether these identified types present a uniform picture in terms of their economic situation during the introduction of the new strategy, or whether we can discover significant differences.

3. Economic situation and different types of strategic change

The term "economic situation" is specified here by the results of the annual balance sheets. The analysis conducted is well aware of all the known shortcomings of such balance sheets (Hauschildt 1996, pp. 1-14; Küting, Weber 1997, pp. 48-54; Leker 1993, pp. 33-80). These shortcomings must, however, be viewed in a relative sense, as almost all comparable studies have also used balance sheets in analyzing economic situation.

For an initial evaluation of the differences between the published and actual corporate performance, we use, along with return on equity and equity ratio, two shareholder-oriented financial ratios (Chaganti, Damanpour 1991, p. 484; Samuels, Brayshaw, Craner 1995, pp. 12-21) and compare them with the solvency rating - as stated in the credit file - for an additional indicator of corporate performance. If the comparison reveals substantial differences, we look for the reasons behind this. If the differences detected are only minor ones, however, we can assume that the description of the economic situation was relatively valid.

First, we study a graphic comparison between the financial ratios of the corporations that underwent a strategic change versus those of a reference group. The reference group is comprised of an aggregate of approximately 18,000 companies differentiated to industry and year of balance sheet by the Federal Bank in its November issue (Deutsche Bundesbank 97-90).

Then, we can more closely analyze the individual characteristics of the four empirically derived types of strategic changes. Due to the small number available for each type, only robust methods of analysis - examining the predominant tendency - are used here. We apply the U-test developed by Mann and Whitney for our comparison of unrelated samples, while testing for the related samples using the Wilcoxon test for pair differences (SPSS Inc. 1997, pp. 317-325; Zöfel 1988, pp. 144-164).

Finally, we look at the solvency ratings for the strategic change companies, comparing these with the results of the balance sheets analysis.

3.1 Analysis of performance

Viewing the development of return on equity for all 65 corporations in comparison with the reference group, it becomes apparent that the situation described at the beginning of this study holds true in the case of our own sample.

First, return on equity of the corporations that have undertaken a strategic change in the year "t" of the period researched are considerably lower than the corresponding values of the reference group. Second, we notice a clear and significantly high deterioration of return on equity in the years prior to the change - whereas the profitability of shareholders' investments shows a slightly improved development following the strategic change.

It is of further interest to note that the reference group also undergoes a moderate reduction in returns during the first four years of the research period. This indicates that the whole industry has been confronted with difficulties during the research period.

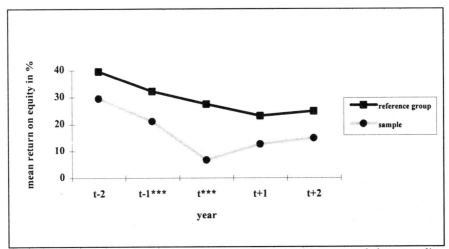

*** $p < .001$, indicates significant difference between this year and the preceding year within the sample using the Wilcoxon-Test

Figure 2: Return on equity

Next, we will try to assess the development of the profits in terms of the four types.

(a) *Median*[†]: Return on equity

	t_{-2}	t_{-1}	t_0	t_1	t_2
Type of change					
1) Expansionist	26,55	23,95	25,54	18,74	19,31
2) Innovator	38,17	16,22 *	8,32 ** 1*	14,27	13,14
3) Reallocator	18,30 2*	18,60	7,13 1**	4,00 1*	10,49
4) Concentrator	23,42	11,97 ***	-2,16 ** 1***	8,39	4,25

[†] N's were respectively 20, 19, 11 and 15 for the groups of change types as listed, because data are not available for all firms in t_{-2} and t_{+2}, the number of observations in these years may vary. * $P < .05$; ** $P < .01$, *** $P < .001$
Median$_i^x$; x = Significant difference between this year and the preceding year within the cluster-type using the Wilcoxon-Test; i = Significant difference between cluster-type i and the respective cluster-type within the analyzed year using the U-Test by Mann/Whitney.

Table 4: Return on equity in different groups of strategic change

It is obvious at first glance that the profit development of these four types is marked by noticeable and significant differences. For instance, we observe in type 1 (Expansionist) a development contrary to the common tendency: the return on equity is overall on quite a high level, even increasing as the year of change approaches. It is, therefore, significantly higher at the time of the change than the other three types. After the change has taken place, however, we find that - again, contrary to the general trend in companies with a strategic change and at the same time corresponding with the profit development of the reference group - the profit actually deteriorates and cannot even be entirely compensated in the second year following the change. For this reason, there is no longer a significant difference between type 1 and the other types at this point in the change process.

Type 2 (Innovator) and type 4 (Concentrator) show a significant decrease in profits in the period prior to change, and, with their dramatic and in both cases outstanding development, make the most distinctive impression from an aggregate viewpoint. Also: the clear increase in post-change profits is found only in these two types, whereby increase in the case of focus on main areas of business (type 4) does not appear to be of a lasting nature.

Type 3 corporations (Reallocator) - where strategic changes are characterized by entry into new purchase markets and the removal of production to low-wage countries - exhibit yet another form of profit development. Evident here is the continued decrease in profits, beginning in the year before the change and enduring beyond the change itself. Only in the final year of the research period could this development be brought to a halt and an increase in return on equity be achieved.

In summary, we find that observing return on equity in terms of the different types of strategic change actually exposes some remarkable differences. Particularly in the cases of type 1 (Expansionist) and type 3 (Reallocator), we see development patterns that differ considerably from the standard results described above.

3.2 Analysis of financial situation

With respect to the financial situation, we again begin with a comparison between those corporations which have undertaken a strategic change and those in the reference group.

In addition to a comparatively clear stability in the equity ratio for both groups, it is also apparent that the reference group's equity ratio is around 8 percentage points lower than that of the change corporations group during the research period. Consequently, we have obtained a result which raises the question of the representative quality of the reference group, while at the same time shedding new light on the differences in the development of the equity profits.

If we keep in mind that the interpretation of return on equity is based on the interworking of the numerator and denominator of a given ratio, we can observe - being aware of the differences found in the equity ratio - that the changing companies' low equity profit throughout the entire research period can be attributed to the comparatively higher and not uniformly profitable equity ratio values. On the other hand, the decrease in return on equity in the strategic

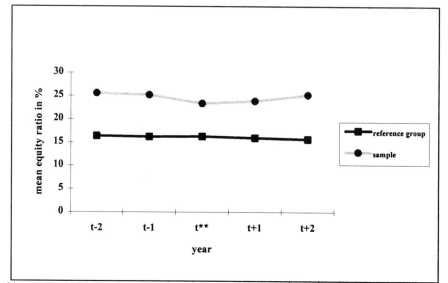

** p < .01, indicates significant difference between this year and the preceding year within the sample using the Wilcoxon-Test.

Figure 3: Equity ratio

change year can be explained by the underproportional success of the company, compared to the changes made in the equity.

In Table 5, arranged by the types of strategic change, we see similar results.

The fact that the equity ratio in the year after the strategic change seems to be almost constant confirms that the different paths of profit development discussed are also attributable to deviant changes in success.

Three further aspects stand out:

1. In at least four out of five years, the expansionist and the reallocator show a significantly higher equity ratio than the innovator and the concentrator - a result that could be useful in a resource-oriented analysis of the various change types (Grant 1991; Mahoney, Pandian 1992; Penrose 1959, pp. 65-87).

2. The expansionist's market-oriented shift leads to a later decrease in the equity ratio.

(a) Median†: **Equity Ratio**

	t_{-2}	t_{-1}	t_0	t_1	t_2
Type of change					
1) Expansionist	30,04	32,12	27,75	28,03	20,50
2) Innovator	19,62 $_{1**}$	16,79 $_{1***}$	17,51 $_{1***}$	15,85 $_{1**}$	19,17
3) Reallocator	27,56	27,52	28,10 $_{2*}$	28,73	32,19
4) Concentrator	20,68 $_{1*}$	18,51 $_{1**}$	15,54 $_{1**}$	17,76 $_{1*}$	22,96

† N`s were respectively 20, 19, 11 and 15 for the groups of change types as listed, because data are not available for all firms in t_{-2} and t_{+2}, the number of observations in these years may vary. * P < .05; ** P < .01, *** P < .001
Median$_i^x$; x = Significant difference between this year and the preceding year within the cluster-type using the Wilcoxon-Test; i = Significant difference between cluster-type i and the respective cluster-type within the analyzed year using the U-Test by Mann/Whitney.

Table 5: Equity ratio in different groups of strategic change

3. The distinctly different, expansive types of strategic change have no visible effects on the specified equity ratio. If we recall the interaction of the numerator and denominator: It appears that companies undertaking a strategic change tend to practice a development in which equity and total capital nearly correspond.

3.3 Analysis of solvency rating

When a bank's credit analysts rate the solvency of a company, this is without a doubt influenced strongly by the balance sheet of the company in question (Bieg 1984, pp. 489-490; Breuer 1991).

Nevertheless, other important so-called qualitative indicators also play a role in such evaluations - for example, judgment of management quality and of opportunities in the markets. These indicators may well be in a position to produce a different result. In addition, the credit analyst's rating includes up-to-date developmental tendencies and can report on current affairs (Hauschildt, Leker 1995, cc. 1323-1335; Leins 1993).

In the following, we will more specifically scrutinize the bank's ratings for the companies which have undergone a strategic change.

– **"Profit"**: a relative estimation of the company profit compared to its competitors based on a 9-point scale, where "1" equals the Standard & Poors issuer credit rating "AAA", and a higher number indicates a deterioration in profit ,

– **"Capital"**: an evaluation of the financial structure on the basis of the equity capital and liable capital available, based on a 9-point scale,

– **"Current"**: an estimate of the economic development since the last balance sheet on the basis of turnover and profits, based on a 9 point scale, and, moreover,

– the aggregated overall evaluation of **"Solvency"** (including the aspects "environment and production risks", "flexibility in adjusting to changes in demand", "management quality", and "flexibility of financial management").

(a) Means[t]: Solvency rating

	t_{-2}	t_{-1}	t_0	t_1	t_2
Ratings					
1) Profit	1,75	2,61 *	3,07	3,50	3,65
2) Capital	1,67	1,88	2,14	2,42	2,85 *
3) Current	3,42	4,00	3,66	3,13 *	3,05
4) Solvency	2,17	2,64	2,68	2,75	2,71

[t] N's were respectively 20, 19, 11 and 15 for the groups of change types as listed, because data are not available for all firms in t_{-2} and t_{-1}, the number of observations in these years may vary. * P < .05
Median^{x}; x = Significant difference between this year and the preceding year within the cluster-type using the Wilcoxon-Test;

Table 6: Solvency ratings for all companies undergoing a strategic change

To begin with, it is conspicuous that the aggregated solvency rating - with respect to the strategic change in year "t" and the profit analysis results - shows only a minor change during the time under observation. After a small decrease at the beginning of the period under analysis, the solvency indicator achieves and maintains a constant value at about 2.5. This corresponds to a good to average solvency. In terms of the Standard & Poor's issuer credit rating system, 2.5 equals a rating between "A" and "BBB". The strategic change seems, therefore, not to have a significant influence on the credit analyst's solvency rating, at least from this overall viewpoint.

A somewhat different picture emerges, however, upon examining the individual component ratings more closely. For instance, the estimations for the components "profit" and "capital" decline steadily throughout the entire analysis period. In particular, the significant decline in the estimate of profits prior to the strategic change corresponds with the results of the balance sheet analysis.

(a) Means[†]: "Current" solvency rating

	t_{-2}	t_{-1}	t_0	t_1	t_2
Type of change					
1) Expansionist	3,00	2,82	2,41 [*]	2,70	2,30
2) Innovator	-	4,86	4,82 [1**]	3,67 [*]	3,32
3) Reallocator	5,00	4,29	3,80	2,55	4,36
4) Concentrator	-	3,90	3,67	3,47	2,73

[†] N's were respectively 20, 19, 11 and 15 for the groups of change types as listed, because data are not available for all firms in t_{-2} and t_{-1}, the number of observations in these years may vary. [*] $P < .05$; [**] $P < .01$
Median[x]; x = Significant difference between this year and the preceding year within the cluster-type using the Wilcoxon-Test; i = Significant difference between cluster-type i and the respective cluster-type within the analyzed year using the U-Test by Mann/Whitney.

Table 7: "Current" solvency rating in different groups of strategic change

We notice, too, that the initially negative estimate of the current turnover and profit development show significant improvement in the change year and the year thereafter. This phenomenon, along with the quality indicators, helps to stabilize the aggregated solvency rating.

In view of the tendencies explained above, it appears appropriate to concentrate at this point on the component estimate "Current", for each of the different types of strategic change.

Except for the type 1 "Expansionist", who enjoys a high instance of agreement between the current turnover estimate/profit situation and the corresponding estimates derived from the balance sheet analysis, we find that the remaining types show certain remarkable discrepancies. If we consider that the analysts generally evaluate their customers with a time lag of nearly one year, and we make allowances for this in the form of some adjustments, the results may indeed come closer; however, there are differences remaining. These differences might be explained to a degree, in that the credit analyst approaches the profit situation with a stronger emphasis on the creditor-oriented view of financial ratios and, moreover, takes qualitative indicators into account.

Despite of this, it is noteworthy that in the up-to-date evaluation of the turnover/profit situation by the credit analysts, a positive rating emerges for the type 2 "Innovator" and the type 4 "Concentrator" following the strategic change.

The originally improved rating for type 3 "Reallocator" in the year after the change, however, is curbed distinctly in the final year, putting this strategic change type back to the low level it occupied prior to the change.

4. Summary

An undifferentiated analysis of companies that have undertaken strategic change leads to an unjustified generalization of the object of analysis. The typology introduced here - four empirically derived types of strategic change - reveals different forms of the progression of economic situation. Each situation is influenced by respective change types. Type 1 "Expansionist", in particular, and also type 3 "Reallocator" show results that differ quite significantly from the other change types.

Also, if we take the solvency ratings of credit analysts into account, we find that the fluctuations demonstrated in the financial ratios are expressed here in a very reduced form: For instance, the deterioration of the financial ratios for type 1 "Expansionist" is seen only in the corresponding evaluation of the solvency component. All in all, the analysts' opinions point to this type as being the one which is not influenced - in terms of economic situation - by a strategic change.

Type 2 "Innovator" and type 4 "Concentrator" show, however, noticeable improvement in their economic situation after the strategic change has been implemented, at least where profitability is concerned.

Type 3, the "Reallocator", can also show financial ratios with an improvement in profitability following the strategic change. Yet these are strongly called into question by a continuously deteriorating profitability evaluation by the credit analysts and a critical estimation of the path future developments will take.

These results ask for further research which would analyze a longer period following the strategic change and also take into account other known strategic success factors (Buzzell, Gale 1987; Hauschildt, Grün 1993).

Appendix

	Definitions of financial ratios:
1) Return on equity =	(net income+ taxes on income / equity) x 100
2) Equity ratio =	(equity / total capital) x 100
with equity =	subscribed capital + capital reserve + revenue reserve - reserve for own shares + retained profits + 0,5 x special item with an equity component - capitalized startup and expansion costs - goodwill - unamortized debt discounts

Evaluation of dynamic technological developments by means of patent data

Holger Ernst

Evaluation of dynamic technological developments by means of patent data

Holger Ernst

1. Introduction

Technological change has been identified to be the major driving force for economic development (Solow 1957). Technological change is mainly determined by public and particularly by private research and development (R&D). Industrial R&D is the most important source for product or process innovations which allows companies to gain competitive advantages leading to sustained economic growth (Brockhoff 1994; Hauschildt 1997). This finding is supported by empirical studies at the firm level which found a positive relationship between R&D expenditures and various measures of commercial success, e.g., measures of productivity, growth and profitability (Mairesse, Sassenou 1991; Morbey, Reithner 1990; Capon et al. 1990).

In the academic literature and the consulting business it is stressed that companies are not well advised if they react to the increasing technological competition only by increasing their total level of R&D expenditures. It is argued that the effective use of scarce R&D resources in those R&D projects which yield the most profound and sustainable advantages over the competition is becoming increasingly important (Brockhoff 1994; Sommerlatte 1995). Thus, various planning instruments have been suggested to support the effective allocation of R&D resources. Among them, different types of technology portfolios have been put forward (Brockhoff 1994; Pfeiffer et al. 1986).

Traditional technology portfolios are mainly based on subjective evaluations of technological positions. However, it has been observed that these evaluations can differ substantially depending on the interviewed experts (Möhrle, Voigt 1993). In addition, this type of portfolio does not take dynamic changes of positions in the portfolio matrix into account. Traditional portfolios only allow for static comparisons at certain points of time. It will be shown in this article that technology portfolios based on patent data - patent portfolios - offer an interesting way to overcome both shortcomings of traditional technology portfolios. They are based on objective measures and further allow the incorporation of dynamic changes of portfolio positions into the strategic R&D decision process.

The use of patent data mainly rests on the assumption that they sufficiently reflect the technological activities of firms. Major support for the use of patents as a measure for R&D outcome comes from quantitative empirical research, where the relationship between R&D and patents at the company level was examined (Bound et al. 1984; Hall et al. 1986; Pakes, Griliches 1984; Scherer 1983). Griliches et al. (1986) summarize this stream of research: "Not only do firms that spend more on R&D receive more patents, but also when a firm changes its R&D expenditures, parallel changes occur in its level of patenting" (Griliches et al. 1986, p. 7). Furthermore, a positive relationship between patent applications and lagged sales growth was found (Ernst 1996). This result proves to be very valuable, since it goes beyond the input-oriented measure of the level of R&D spending and supports the use of patent data even as an output measure of R&D, since patents indicate those technological activities which lead to subsequent market changes (Griliches 1990). The output impact of R&D is best mirrored by those patents, which are of higher technological and commercial quality than an average patent application. Patents granted, valid patents, international patent applications and patent citations have frequently been identified as quality signs of patents (Albert et al. 1991; Basberg 1987; Ernst 1995; Harhoff et al. 1997; Narin et al. 1987). Including these indicators of patenting quality in patent portfolios enhances its meaningfulness.

This paper is organized as follows. In section 2, we will first briefly discuss the structure of traditional technology portfolios and will then, more extensively, describe the structure of patent portfolios. In section 3 we will illustrate the use of patent portfolios by applying this method to a group of international companies from the chemical industry. Here, we will analyze dynamic variations of portfolio positions and their implications. We conclude this paper with a brief summary and suggestions for further research in section 4.

2. The portfolio method for strategic R&D planning

2.1 Traditional technology portfolios

A variety of portfolio matrices has been developed to support the effective allocation of scarce R&D resources to specific technological fields. Brockhoff distinguishes between marketing- and technology-dominated approaches and attempts to integrate both views in one portfolio matrix (Brockhoff 1994). In marketing-dominated portfolios, strategic R&D investment decisions are directly derived from the product positions shown in the market portfolio. Different technologies which lie or could lie behind these products are not explicitly considered. Thus, the attractiveness of single technologies cannot be evaluated. This approach is problematic because new emerging technologies which lead to the obsolescence of former highly attractive products may be overlooked (Pfeiffer et al. 1986). Furthermore, it is argued that the predominant market orientation of R&D may lead to marginal improvements only, and substantial innovations are not realized (Brockhoff 1985).

In contrast, technology-dominated portfolios allow the direct evaluation of product or process technologies. The general structure of technology portfolios is illustrated by using the portfolio matrix suggested by Pfeiffer et al. (1986, 1991, 1995).[1]

The technology portfolio shows the typical characteristics of two-dimensional portfolio matrices. On the abscissa, an indicator of a company's capabilities or strengths in a specific technological field is displayed. This value is predominately determined by the behavior of the firm under consideration. Pfeiffer et al. suggest measuring resource strength per technological field as a multidimensional construct which basically consists of two elements: a company's know-how and its financial strengths in order to build new know-how. These two elements are further broken down to separate items which are subjectively measured on a five-point rating scale (Pfeiffer et al. 1991).

---

[1] An overview of other technology portfolios is given by Brockhoff (1994).

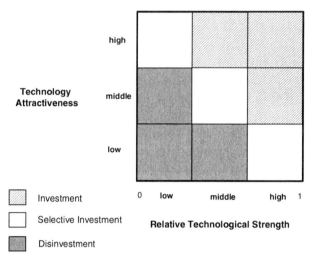

Source: Pfeiffer et al., 1986, p. 122.

Figure 1: Structure of technology portfolios

On the ordinate, an indicator of a technology's attractiveness is displayed. This value is predominately determined by the behavior of all other firms being active in the respective technological field. Technology attractiveness is also measured as a multidimensional construct which mainly measures a technology's development potential and its market potential. Again, corresponding single items are to be assessed by experts on the same five-point scale (Pfeiffer et al. 1991). Based on the position of technologies in the portfolio matrix, the pursuit of general R&D investment strategies is recommended. Basically, a company should invest in technologies positioned in the upper right part of Figure 1 as these technologies are attractive and the company holds strong positions. In contrast, the company should refrain from investments in those technologies positioned in the lower left part of Figure 1. For the diagonal, selective R&D investment decisions are recommended.

Technology portfolios are subject to criticism known from other portfolio matrices (Brockhoff 1993; 1994). Here, we would like to point out two aspects which are of relevance to this paper.

First, traditional technology portfolios are based on subjective personal judgments. It can be assumed that the assessment of both portfolio dimensions vary according to the interviewed person. It is known from organizational and marketing research that even knowledgeable informants may disagree because they hold different organizational positions and, thus, different perspectives on the same organizational phenomena. Here, a so-called informant bias is present which taints respondents' reports (Kumar et al. 1993). In particular, the assessment of a technology's attractiveness to a company requires the integration of the major functional departments (Marketing, R&D, Production). However, it was shown that interface problems between Marketing and R&D occur especially in the screening phase for new products (technologies) which leads to an informant bias (Ernst, Teichert 1998). Thus, the assessment of a technology's development and market potential may not be agreed upon by respondents from Marketing and R&D. Consequently, positions in technology portfolios differ with respect to the interviewed person. Systematic research on respondent effects on the validity of technology portfolio positions is not (yet) available. However, observations made during a practical application of a similar technology portfolio method give further hints. Experts did not always have the knowledge to assess all the technological items they were asked, and they substantially disagreed in their individual assessments (Möhrle, Voigt 1993).[2] In sum, technology portfolios have to rely on personal judgments which lead to substantial measurement problems. They can partially be avoided by using multiple and knowledgeable respondents. However, aggregating different answers to one single construct for both portfolio dimensions continues to be problematic.

Second, technology portfolios mirror technological positions at a certain point of time. However, they fail to show dynamic changes of portfolio positions. These dynamic developments could have an impact on the formulation of present R&D strategies. If a company decides to use a technology portfolio to support its present R&D investment decisions, it will get a static picture of today's technological positions. However, it might be of importance to get further information on the evolution of these positions over the preceding

---

[2] E.g., a median value of three could be observed for the deviating answers on a ten point scale. Furthermore, 17% of the answers deviated more than five points.

years. Questions to be answered include, for example: How has the attractiveness of a particular technological field changed in recent years? In what phase of the technological life cycle is a technology? Is a technology about to be rejuvenated? Has the technological emphasis of competitors changed over the years? Have our capabilities in specific technologies increased or decreased? Answers to these questions can be given only, if the history of portfolio positions is incorporated in technology portfolio illustrations. However, traditional technology portfolios do not provide this information.

We will argue in the following that technology portfolios based on less but objective patent data may offer an interesting solution to both problems, subjectivity and missing dynamism, inherent in traditional technology portfolios.

2.2 A technology portfolio based on patent data

The patent portfolio has the same basic structure as the traditional technology portfolios described above. A measure of technological strength is used for the abscissa and a measure of technology attractiveness is used for the ordinate. However, both basic dimensions of technology portfolios are now assessed by means of patent data. The basic concept of a patent portfolio was first introduced by Brockhoff (1992). Since then, the contents of patent portfolios have been expanded and practically applied on a large scale (Ernst 1998). Figure 2 illustrates the basic structure of a patent portfolio.

(1) On the abscissa, we measure *a company's patent position in a specific technological field relative to its competitors*. The relative patent position per company and technological field (RPP$_{if}$) is defined as follows:

(1) $RPP_{if} = PP_{if} / PP_{i*f}$

(1a) $PP_{if} = PA_{if} \circ PQ_{if}$

(1b) $PQ_{if} = \sum_{k=1}^{K} RPQ_{ifk}$

(1c) $RPQ_{ifk} = QI_{ifk} / \left[(\sum_{i=1}^{I} QI_{ifk}) / I \right]$

with the following parameters:

PA = Number of Patent Applications; PP = Patent Position; PQ = Patenting Quality; QI = Indicator of Patenting Quality; i = Company (i = 1,..,I); i* = Company with strongest Patent Position (Benchmark); k = Indicator of Patenting Quality (k = 1,..,K); f = Technological Field.

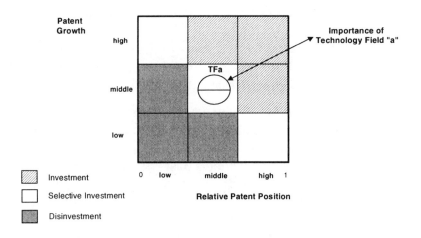

Figure 2: Structure of patent portfolios

The number of patent applications (PA) reflects the patenting activity of companies. It shows the extent of R&D activities carried out in specific technological fields and further demonstrates the patentee's intention to use the invention economically in the market. However, by using patent applications only, one cannot distinguish between diverging qualities of inventions. Several indicators for patenting quality (QI) can be used:

1. **Rate of patents granted**: A patent will be granted only, if the invention consists of new technological elements (e.g., § 1(1) German Patent Law). Therefore, a patent granted is believed to be of higher technological value than the mere patent application (Basberg 1987). The rate of patents granted can differ substantially between companies and thus possibly deflates a high number of patent applications (Brockhoff 1992). The rate of patents granted is measured as the number of patents granted over patent

applications minus patent applications under examination.[3] The rate of patents granted is considered as an indicator of technological patenting quality (Ernst 1996).

2. **Rate of valid patents**: Patents are valid if they have been previously granted and the protection fee is still paid by the patentee. Assuming rational decision behavior it can be reasoned that valid patents are still economically valuable for the company, i.e. the economic benefit is larger than the cost to maintain the patent. It can further be assumed that the share of valid patents correlates with the renewal time of a patent, which is widely believed to be an indicator of high-quality patents (Schankerman 1991; Schankerman, Pakes 1986). The rate of valid patents is measured as the number of valid patents over patents granted (Ernst 1995).

3. **Share of international patent applications**: International patent applications are considered to be more valuable, since the cost of obtaining an international patent is substantially higher than that of a national patent application (Basberg 1987; Griliches 1990; Schmoch 1990). In the literature, patent applications in the US have been frequently used to measure patenting quality (Basberg 1983; Glismann, Horn 1988; Soete 1987). For example, a company's share of US patents could be measured as the quotient of its US patents and the total number of its patent applications. In general, the selection of an appropriate foreign country to be used depends on the country of origin of the companies under investigation. It was shown for German companies that their share of US patents correlates with commercial success (Ernst 1996). If international comparisons are made, different indicators of international patent applications are to be used, e.g., if German companies are to be evaluated against competitors from the US, the use of US patents as an indicator of patenting quality is misleading. In these cases, it is advisable to either use patent applications at a foreign neutral third patent office (Pavitt 1988), e.g., the European Patent Office (EPO) or to define a new indicator of international patenting activity tailor-made to fit the requirements of the respective investigation.

---

[3] Patent applications under examination are subtracted from the total number of patent applications since they can still be granted in the future (Ernst 1995).

4. **Patent citation ratio**: Patents are used by patent inspectors at the patent office to document the state of technology when they check if a patent application contains new features which go beyond what has been known so far. This procedure leads to patent citations. The number of citations received by a patent in subsequent patent documents is often interpreted as a sign of an economically important invention (Albert et al. 1991; Carpenter et al. 1981; Harhoff et al. 1997). The average citation ratio can be computed by dividing the number of citations by the total number of patent applications (Narin 1987).

The last three indicators are of special importance for determining a patent's quality. Several empirical studies prove that these quality indicators positively correlate with measures of commercial success on the company level (Ernst 1995; Narin 1987). Thus, they are also referred to as indicators of the economical patenting quality (Ernst 1996). The same correlation could not be proven true for the share of patents granted as a measure of technological patenting quality (Ernst 1995). Thus, as it is known from market portfolios, where a company's relative market share as the major driver of company performance (Schoeffler et al. 1974) is displayed on the abscissa, the inclusion of success-related quality measures of patents does enhance the meaningfulness of technological positions displayed in patent portfolios.

It has been argued that the reliance on only one measure of patenting quality may be subject to significant evaluation failures caused by varying values for different quality indicators. Therefore, we aggregate the indicators of patenting quality described above to a construct of overall patenting quality, which is believed to provide a more stable assessment of companies' patenting quality (Ernst 1996):

5. **Total patenting quality**: A construct of patenting quality per technological field (PQ_{if}) is used which consists of the sum of relative measures for each individual indicator of patenting quality (RPQ_{ifk}). Relative values are calculated by relating the respective indicator of patenting quality for each company to its mean value over all companies under consideration (Ernst 1996; see formula (1c)). It should be pointed out that the number or content of the various indicators of patenting quality to be included in the construct of patenting quality can be subject to variations in accordance with the specific objectives of patent portfolio illustrations.

Finally, the relative patent position of each company per techno-logical field (RPP$_{if}$) is measured as the quotient of its patent position in a technological field (PP$_{if}$) and the strongest patent position of any of the relevant competitors in the respective technological field (PP$_{i*f}$). Thus, the maximum value for RPP$_{i*f}$ is one. Using the strong-est patent position per technological field as a benchmark allows the direct identification of leading and following companies in specific technological fields and illustrates immediately the distances be-tween the respective companies (Ernst 1998).

(2) On the ordinate, we measure the attractiveness of each tech-nological field by using *growth rates of patent applications*. Brock-hoff, for example, suggests measuring the growth of patent applica-tions in a specific technological field during the past 4 years relative to the growth in the preceding 16 years, which covers the 20-year patenting period (Brockhoff 1992). In the literature, numerous other growth measures can be found (Ernst 1998; Faust 1989; Marmor et al. 1979). Most of these measures of patent growth stress recent changes in patent growth, i.e., high growth rates in recent years rela-tive to patent growth in preceding years are interpreted as an indi-cator of high technology attractiveness. Furthermore, it has been suggested that one should measure relative growth rates (Ernst 1998), e.g., by dividing the growth rate of patents in a technological field by the average growth rate of all other technological fields un-der investigation in the patent portfolios. Both, the respective refer-ence measure to calculate relative rates of patent growth and the aforementioned definition of time intervals can vary between differ-ent portfolios. It was shown that different measures of patent growth lead to diverging portfolio positions, which has an impact on the conclusions drawn from the portfolios (Ernst 1998). In principle, relative patent growth in technological field f (RPG$_f$) can be meas-ured as follows:[4]

(2) $\quad RPG_f = PG_f / PG_R$

(2a) $\quad PG_f = PG_{ft-\tau} / PG_{ft-\theta}$

(2b) $\quad PG_R = PG_{Rt-\tau} / PG_{Rt-\theta}$

---

[4] The natural logarithm is used to measure yearly growth rates correctly. The arithmetical mean is used for averaging growth rates. The validity of this proce-dure to calculate growth rates was proved (Wetzel 1964).

with the following parameters:

PG = Patent Growth; t = Reference Point for Time Interval;
τ = Time Lag for Numerator (Recent Patent Growth);
θ = Time Lag for Denominator (Preceding Patent Growth);
f = Technological Field; R = Reference to Measure Relative Growth Rates.

The underlying assumption of using patent growth rates as a measure of technology attractiveness is that high patenting activity, caused either by a higher number of patent applications per company and/or by an increasing number of patentees entering a specific technological field, reflects the attractiveness of technologies and, beyond that, even the attractiveness of market opportunities. The following reasons support this assumption. It can first be argued that a patent application shows a patentee's willingness to market the invention, which in turn shows that the respective company assumes a commercial opportunity for its invention (Griliches 1990). Second, empirical research on the company level shows that patent positions are related to commercial performance for a cross-section of firms (Ernst 1995) and, even, for cross-section time-series data, that patent applications lead to subsequent sales increases, indicating a causal relationship between patenting activity and market changes (Ernst 1996). Third, a substantial number of studies on the level of technological fields have found an almost parallel development of patenting activity and market growth of products based on the underlying technologies (Achilladelis 1993; Ernst 1997).[5]

(3) The circle size of the technological fields displayed in patent portfolios reflects the *distribution of total company's patents among technological fields*. This indicates the importance of each technology within the company's R&D portfolio. Technology importance given to technological field f by company i (TI_{if}) is calculated by the number of patent applications in a technological field (PA_{if}) relative to the total number of patent applications of the company (PA_i) in question:

---

[5] Patent statistics published by the German Patent Office (GPO) show the increasing number of patent applications in the technological or product field air-bag in the 1980s (GPO 1994). During this time, airbags have become standard even in mid-sized and small cars. Similar developments can be observed for anti-blocking-systems in the automobile industry and mobile phones in the telecommunication industry.

(3) $TI_{if} = PA_{if} / PA_i$

with the following parameters:

TI = Technology Importance; PA = Patent Applications;
i = Company (i,..,I); f = Technological Field.

In general, the patent portfolio can be used to evaluate technological strengths and weaknesses of competing companies with respect to different technological fields. This information supports strategic R&D investment decisions. As we know from traditional technology portfolios, companies should invest in growing technological fields where they hold strong patent positions, whereas they should disinvest in low-growing (declining) technologies where they hold rather weak patent positions (see Figure 2).

At the beginning of section 2.1, it was explained that traditional technology portfolios and patent portfolios are technology-oriented tools for strategic R&D planning. Both need to be aligned with other strategic planning instruments in order to avoid one-sided, technology-dominated misconceptions. Various methods to integrate traditional technology portfolios (Benkenstein 1989; Brockhoff 1994) and patent portfolios with market-portfolios have been suggested (Ernst 1996).

In the following, we will illustrate for a group of international companies from the chemical industry dynamic position changes in patent portfolios and their impact on strategic R&D decision making. Other than traditional technology portfolios, the patent portfolio method allows the analysis dynamic aspects of portfolio positions. In other words, the emerging of today's positions in patent portfolios over preceding years can be made visible because patent data can easily be assigned to their time of origin. Knowing the dynamic evolvement of patent positions over time adds valuable information to the interpretation of portfolio positions.

3. Application of the dynamic patent portfolio method in the chemical industry

3.1 Preliminary work

We applied the patent portfolio for a group of seven major international companies operating in a specific segment of the chemical industry. The companies came from Germany, Japan, and the US. For reasons of confidentiality we cannot report any information which could lead to the identification of the companies under investigation. During this case study we closely co-operated with senior R&D managers from the German companies. During an initial workshop, five technological fields were identified to be included in the patent portfolio analyses.

It was decided to take patent applications that assure legal protection for an invention on German territory as the basis for our analyses. These patent applications can either be direct patent applications at the German Patent Office (GPO), or patent applications via the Patent Cooperation Treaty (PCT) mechanism or via the EPO. The major reason for this approach was our intention to measure the quality of patents, e.g., by means of patents granted and valid patents. Since patents are territorial rights, these indicators of legal status can only be compared at the same patent office. Here, the German companies were mainly interested in considering the patent situation on their local market. Furthermore, Germany can be considered as the most important market in Europe, especially for the chemical products under consideration. In general, 90% of patent applications at the EPO claim Germany as a designated state (Schmoch et al. 1988). Thus, patent applications at the GPO provide an almost complete picture of German and European patenting activities. In addition, we found for our sample that all European patent applications were also filed at the GPO. Senior R&D managers were convinced that their competitors would also file all relevant patents at the GPO. However, it has to be maintained that patenting activity of German companies may be overestimated due to their home-country advantage, whereas patenting quality for their foreign

competitors may be exaggerated because international patent applications are filed more selectively (Ernst 1998).

Patent data was derived from the patent databases PATDPA, WPI
and ESPACE Bulletin from 1978 to date. Since this analysis was
carried out at the end of 1997, patents applied for until the beginning
of 1996 could be considered due to the 18 months time lag between
priority date and publication of the patent application. These patents
had to be allocated to the five technological fields. This can either be
done manually or automatically (Brockhoff 1992). It was of interest
to us to examine the deviation between the two procedures (Ernst
1998). Thus, in one of the companies patents were allocated to the
defined technological fields by their own R&D staff. Patents were
automatically allocated by using a combination of relevant IPC
classes and keywords to define each technological field (Schmoch
1990). In total, 239 patents had to be allocated. Of these, 201 patents
were assigned to the same technological field; eight of the remaining
28 patents were only assigned to a technological field automatically
because they remained unclassified during the manual allocation
procedure. The rather small deviation of 8% of the patents supports
the use of automatic allocation procedures, after a technological
field is properly defined by a combination of IPC classes and precise
technical expressions. Here, the alignment of those who perform the
patent search and experts from R&D is crucial (Ernst 1996). Senior
R&D managers argued that the manual inspection of single patent
documents had always been very burdensome, especially when large
amounts of patent data had to be evaluated and continuously up-
dated. Thus, automatic allocation schemes would greatly enhance
the usefulness of patent portfolios.[6]

---

[6] The manual allocation of patents to technological fields is subject to a bias, if
different persons are involved. Thus, criteria need to be specified which make
the systematic and comparable assignment of patents possible. These criteria
can be used as keywords for automatic allocation.

3.2 Drawing of patent portfolios

The patent portfolios followed the basic structure outlined in section 2.2. However, two remarks concerning the measurement of the two portfolio dimensions need to be made.

1. The relative patent position was measured according to formula (1). The share of patents granted, the share of valid patents, the citation ratio and the share of international patents were used as quality indicators (k) of the patent applications. The share of international patents was measured by dividing a so-called "triad patent" by the total number of patent applications. A "triad patent" is simultaneously filed in Germany, Japan and the US. This was viewed to be an appropriate measure because all of the considered companies operate on a global scale and the use of "triad patents" allows the comparison between companies of different national origin.

2. The total time period was divided into three subperiods in order to analyze dynamic developments of positions in the patent portfolios. Thus, patent portfolios covering the years 1978 to 1985, 1978 to 1990 and 1978 to 1995/6 (date of data retrieval) were constructed which display the technological positions which we would have found if we had drawn single patent portfolios for the years 1985, 1990 and 1995/6. Rates of relative patent growth as an indicator of technology attractiveness were measured according to the basic formula (2) outlined in section 2.2, i.e., recent patent growth for a technological field was divided by recent patent growth in all five technological fields.[7]

---

[7] For example, patent growth in technological field 1 (TF1) in patent portfolio II (1978-1990) is calculated as follows: a) Patent growth (TF1, 1985-1990) / Patent growth (TF1, 1978-1984); b) Patent growth (all five technological fields, 1985-1990) / Patent growth (all five technological fields, 1978-1984); c) Relative Patent growth for TF1 in patent portfolio II (1978-1990) = a/b.

3.3 Analyzing patent portfolio positions

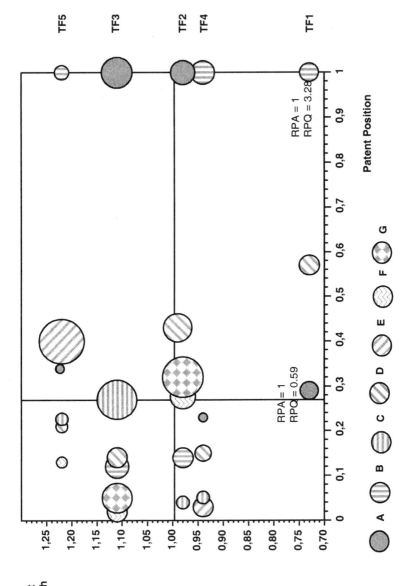

Figure 3: Patent portfolio (1978-1985)

Figure 3 displays the first patent portfolio for the year 1985 covering the period from 1978 to 1985. TF5 and TF3 are the fastest growing technological fields; TF1 is the slowest growing technological field. All technological fields are dominated by either company A or company B. Company A has strong patent positions in TF3 and TF2; it has weak patent positions in the technological fields TF5 and TF4. It seems that company A neglected these two technological fields at that time. In contrast, company B shows an almost opposite pattern: strong patent positions in TF4 and TF5 and weak patent positions in TF2 and TF3. Interestingly, company B has the highest patenting quality in both TF4 and TF5. Obviously, company B had an early major focus on these two technologies. Company B further holds the strongest patent position in TF1. Company A filed the same number of patents in that period (RPA = 1)[8]; however, the quality of its patents (RPQ = 0.59) is much lower than for company B (RPQ = 3.28). Hence, the inclusion of quality indicators for patents can substantially change the relative patent position in patent portfolios (see section 2.2).

Figure 4 displays the second patent portfolio for the year 1990, covering the period from 1978 to 1990. Patent growth in TF4 and TF5 has further increased and is fastest among all technological fields. TF1 and TF2 are least attractive. All technological fields are further dominated by either company A or B. Only company D in TF1 and company C in TF5 come close in particular technological fields. It is striking that company A has dramatically improved its patent position in TF5 and TF4. Company A is the most active patentee in both technological fields in that period; however, patenting quality is lower than for company B, which is still leading in these technological fields. Company A remains the dominant company in the attractive technological field TF3. Company B has the strongest patent positions in TF4 and TF5. In addition, it puts more emphasis on TF4 than company A, and the quality of patents in TF4 and TF5 remains very high. In contrast, company B does not focus on TF2 and lies far behind company A in TF3. Finally, company B lost its dominant patent position in TF1. In that period, company A had closed the gap and even passed company B to gain the strongest patent position in TF1.

---

[8] RPA is measured as described in formula (1). However, patenting quality is not taken into account.

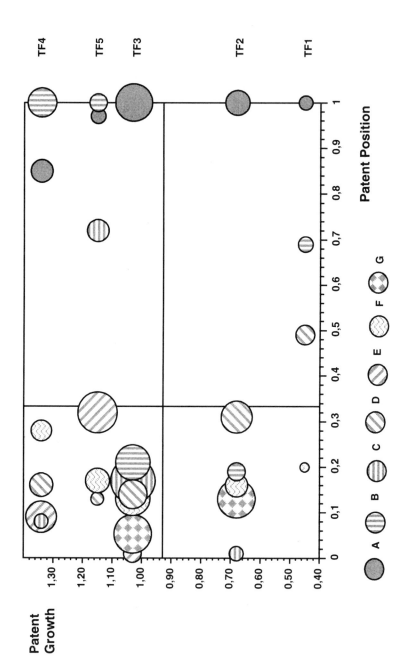

Figure 4: Patent portfolio (1978-1990)

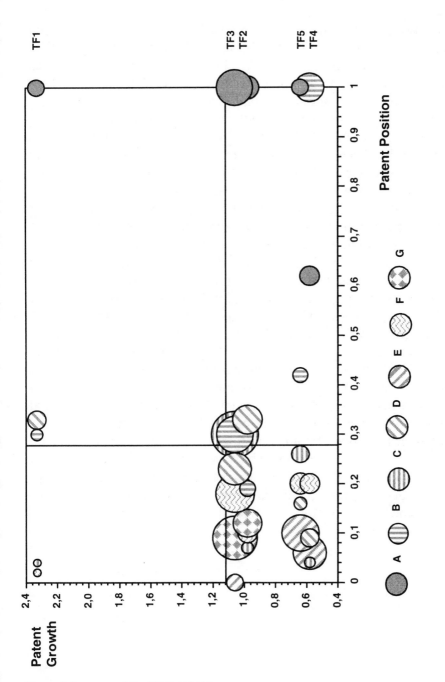

Figure 5: Patent portfolio (1978-1995/6)

Figure 5 displays the third patent portfolio for the year 1995/6, covering the period from 1978 to 1995/6. Surprisingly, patent growth is fastest in those technological fields, TF1 and TF2, which had experienced prior low growth rates. In contrast, patent growth is lowest in the former fast growing TF4 and TF5. Patent growth has been high for TF3 during the entire time period. This seems to be a key technology in the industry under consideration. All technological fields are still dominated by either company A or company B.

The patent portfolio illustration in Figure 5 can further be improved if those patents are excluded from the analysis which had not been renewed during the considered time interval. Thus, we use the patent stock to measure the relative patent position. The patent stock consists of valid (granted and in force) patents and recent patent applications which are still under examination by the patent office (Ernst 1998). In this case, the quality of the patent stock is measured by the remaining two indicators of patenting quality, i.e., the share of international patents and the citation ratio (see section 3.2). The resulting patent portfolio is displayed in Figure 6.

A few differences from Figure 5 are apparent. First, companies D and B get closer to the still leading company A in TF1. Second, company D emerges as the second strongest competitor of company A in TF2. Third, company C and particularly company B close the gap to company A in TF5. Fourth, the leading position of company B in TF4 becomes even more evident. In sum, it should be stressed that this type of portfolio matrix proves to be most valuable if present patent positions of companies are to be assessed. The inclusion of former patent applications which have not been renewed would alter the portfolio illustrations. However, they prove to be very helpful in order to illustrate the evolution of present portfolio positions and should, therefore, not be neglected.

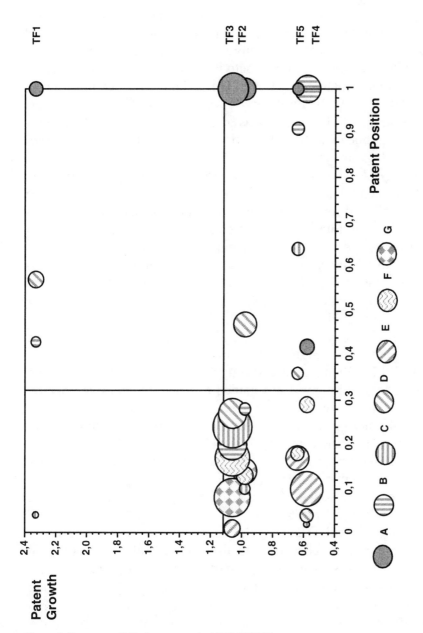

Figure 6: Patent portfolio (patent stock, 1978-1995/6)

3.4 Overall implications of patent portfolios

Recent patent growth is fastest in TF1, TF2, and TF3; recent patent growth is lowest in TF4 and TF5. Future R&D effort should be directed towards those technologies that may promise more sustained competitive advantages than improving other technologies. Company A early focused its R&D activities on TF2 and TF3, where it had a dominant patent position (core technological competencies) from 1978 to 1995/6. Furthermore, company A has the strongest patent position in TF1, which was gained after the year 1985. Thus, company A appears to be well positioned, today, in the most attractive technological fields.

Company A shows an overall strong patenting performance. Across all technological fields, it has been the most active patentee and the quality of its patents is high. Company A takes the strongest patent position in four out of five technological fields and the second strongest patent position in the remaining technological field, TF4. However, the distance to the leading company B is large in TF4. It became obvious from the early patent portfolios that company A did not pay much attention to TF4 and TF5. Here, it followed other companies, especially company B.

Company B has a clear focus on TF4 and TF5. It was among the first companies to recognize the importance of these technological fields and it has maintained a leading position from 1978 to 1995/6. It appears that company B has gained a strong competitive position in this area. In particular, TF4 is given much emphasis, which contrasts to the activities of company A. Furthermore, it has weaker patent positions than company A in TF3, TF2 and TF1. However, the quality of company B's patents is high in the last two technological fields mentioned. In sum, the patent portfolios show a distinct strategic difference of R&D activities between companies A and B from 1978 to 1995/6.

Companies C and D appear as the most serious competitors of companies A and B. Their patents are of high quality and cover all technological fields. Company D's particular strengths can be found in TF1 and TF2, where it has the second strongest patent position behind company A. Company C has a strong patent position (high quality of patents) in TF5 and has a particular R&D emphasis on TF3.

4. Discussion

We successfully applied the patent portfolio method for seven international competitors from the chemical industry. Final discussions with our partners in industry about the results of the study revealed that this instrument was viewed as adding valuable information to the strategic R&D planning process and that, under cost-benefit considerations, it appeared very appealing to practitioners. The cost of patent data retrieval amounted to approximately $6,000, and it took a one-day workshop with senior R&D managers to define the technological fields. However, it has to be added that the costs of patent data analysis cannot be generalized, since they depend on the extent of patent searches performed and the search expertise of the analyst. However, patent data retrieval costs are relatively low compared to other forms of technological information sources (Ashton et al. 1991) and, more important, they have to be judged according to their benefits for strategic R&D planning. Moreover, all managers from the companies involved stressed the advantages of getting an objective picture of technological positions which could further be used to better communicate technological strengths to senior management and/or outside stakeholders like investors or potential partners.

The positive experiences made during this study lead to the insight that the continuous and strategic analysis of patent information should become an essential part of strategic planning activities within each company (Ashton, Sen 1988). This could help to improve the insufficient level of information about competitors' R&D strategies. As the experience from this study further shows, the effective use of patent information in strategic planning can only be achieved by the inclusion of company expertise in the patent retrieval and analysis process. Thus, the outsourcing of strategic patent data analyses to external information brokers does not seem advisable (Reiche, Selzer 1995). Hence, companies need to establish a particular unit, within the organization, which is responsible for the continuous and systematic evaluation of patent information (Ernst 1996).

The study further revealed that, in contrast to traditional technology portfolios, patent portfolios can be used to analyse dynamic technological developments. The emerging of today's positions in patent portfolios over preceding years can be made visible ex post, because patent data can easily be assigned to their time of origin.

Knowing the dynamic evolvement of patent positions over time adds valuable information to the interpretation of portfolio positions. In our study, we were able to analyze dynamic changes with respect to the attractiveness of technological fields, the relative patent position of companies in these technological fields, and the importance given to each technological field by each company. Here, we would like to illustrate this point by giving some examples.

TF3 shows high rates of patent growth over the total time period. It can thus be regarded as a core technology in the respective industry. Companies may not be well advised to neglect this technology. Furthermore, TF5 and TF4 were the fastest growing technological fields until 1990. Since then, TF2 and TF1 have been the fastest growing technological fields, which yields important implications for future R&D investment decisions. To one of the German companies this result came as a surprise, because senior management had been convinced that TF4 and, especially, TF5 would play an important role in the future. It was not perceived before that patent growth had already been high in these technological fields almost ten years previously. In this context, a major difference between companies A and B over time became obvious. Company B was among the first to realize the importance of TF4 and TF5, whereas company A caught up to company B only some years later. In this context, the patent position of company E is of interest. Company E holds patents only in TF4 and TF5. It was also among the first to file patents in these two technological fields (see Figure 3). The latest portfolios (see Figures 5 and 6) show that company E still holds these patents. Hence, these patents must be of substantial importance to company E. Company E may not be considered as a direct competitor of the other six companies since it does not operate along the entire value chain which is typical for this segment of the chemical industry. It rather focuses on the application of certain technologies to develop specific products. Looking at the portfolio may thus reveal that company E is fast in recognizing newly evolving application trends. It may thus be a candidate for consideration by the other chemical companies as a lead user (von Hippel 1986). In sum, these examples and the detailed analyses in sections 3.2 and 3.3 show that dynamic changes in patent portfolio positions over time add valuable information to strategic R&D decision making which would not have been available if a traditional technology portfolio had been drawn up in that industry in 1995/6.

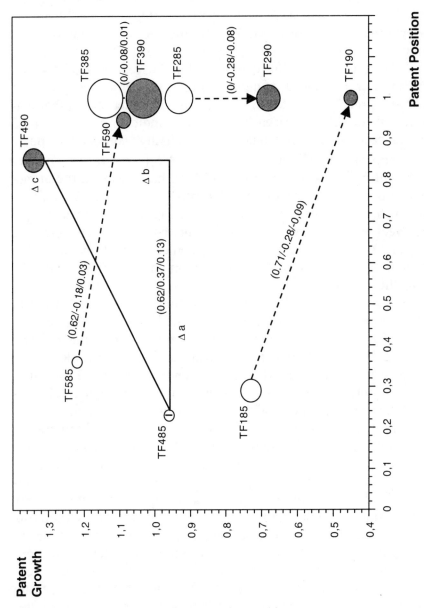

Figure 7: Dynamic development of patent portfolio positions

Dynamic changes in patent portfolios can be made more explicit. Figure 7 illustrates a different approach to analyze dynamic developments of patent portfolio positions. Here, for example, we consider changes of portfolio positions of company A between 1985 and 1990. Dynamic changes are captured by a vector showing the degree of deviation of the three portfolio variables between the two years. Company A had a weak patent position in TF4 in 1985, where it was 0.8 away from the leading company. By the year 1990, however, company A had reduced its deficit by 0.62 and almost caught up to the most strongest patentee in TF4. Figure 7 further shows that patent growth in TF4 had risen by 0.37, thus indicating the increased importance of this technological field. Finally, company A put a larger emphasis on TF4 in 1990 than in 1985 ($\Delta c = 0.13$). This development indicates that company A must have realized the attractiveness of TF4 and, therefore, substantially increased its R&D efforts in this field, resulting in a much stronger patent position. All other vectors in Figure 7 can be interpreted accordingly. Overall, company A remained the strongest patentee in TF2 and TF3 and substantially improved its patent position in TF1, TF4, and TF5 during this period.

This paper also yields some implications for further research. Here, one major aspect ought to be outlined. Since a positive relationship between quality indicators of patents and company performance has been found in empirical studies, we included these measures in the patent portfolios in order to enhance the meaningfulness of portfolio positions. Thus, we indirectly assume a relationship between portfolio positions and company performance. It seems, however, worthwhile to directly test the relationship between a company's position in the patent portfolios and various measures of economic performance. The hypothesis to be tested is that companies holding strong patent positions in highly attractive technological fields are more successful than competitors holding weak patent positions in unattractive technological fields. This hypothesis could be modified to state that companies which first hold strong patent positions in highly attractive technological fields are more successful than those competitors which later achieve strong patent positions in the same technological fields. Both hypotheses can be tested either in a cross-section or a panel analysis, whereby the latter allows the incorporation of lag structures between independent and dependent variables.

Innovation objectives, managerial education and firm performance – an exploratory analysis

Dietmar Harhoff

Innovation objectives, managerial education and firm performance - an exploratory analysis

Dietmar Harhoff

1. Introduction

Personal intuition and casual observation of the business press suggest that people matter. The characteristics of top managers, e.g., their educational background, their career paths, and their personal traits, play some role in the assessment of a firm's prospects. This does not just reflect the demand for managerial folklore. Some narrative accounts in the management literature, e.g., in entrepreneurship, also attest to the importance of individuals (see, e.g., Roberts 1991). Historical studies have linked firm strategy, performance, and idiosyncratic characteristics of top executives - Chandler's (1962) study of organizational structure and strategy being a particularly well-known example. Thus, one cannot simply ignore the potential relationship between a firm's performance and the talents, skills, and training of its top managers. Yet, this nexus has played only a minor role in the literature on strategic management or organization[1].

Is there really a systematic and statistically reliable relationship between a top manager's knowledge and experience and the firm's strategic choices? And why should we expect performance differentials to arise between firms whose top managers have experienced different types of education and career experiences? These two questions are difficult ones, both from a conceptual and an empirical point of view. The main conceptual problem is the simultaneity between firm strategy and top executive characteristics. While top managers will have some impact on strategy, the selection of par-

---

[1] Schrader (1995a) convincingly criticizes various bodies of literature for this neglect. He argues that the heterogeneity of top executives may be a major determinant of the heterogeneity of firm strategies and performance outcomes.

ticular individuals for top management jobs will also be guided by the potential fit between the manager's characteristics and job requirements (the latter again being driven by the firm's strategy). Empirically, it is difficult to observe the overall process during which individuals are selected and then affect firm strategy. Yet, even partial answers to the two questions would appear helpful. This paper attempts to add to the existing literature, presenting a mainly empirical analysis for a sample of West German firms. In many ways, the paper follows the detailed and insightful study by Schrader (1995a), who analyzed the educational and career background of U.S. CEOs and the relationship between CEO characteristics and firm behavior and performance.

The logic of the argument pursued here is as follows: specific forms of education and the experiences made on subsequent career paths endow an executive with particular skills and information. These may give the individual comparative advantages, some of which may be of importance only under particular circumstances and business conditions. This form of comparative advantage may be strong enough to affect the performance of the firm as a whole.

In order for this logic to work, a number of inefficiencies in delegation and top management selection have to be present. The characteristics and abilities of top managers will matter if decision-making and information gathering cannot be delegated perfectly, e.g. in the presence of asymmetric information. Moreover, for performance differentials between companies to show up, the hiring and training of new top managers must be a time-consuming task which prevents firms from replicating situation-contingent top management skills quickly. This paper argues that both conditions are likely to be met in real-world organizations, and that they have important repercussions. The link between characteristics of top executives, strategy and performance is possibly the most important of these implications.

Schrader (1995a) provides comprehensive evidence that there are statistically significant relationships between CEO education, career paths, firm strategy and firm performance. He first demonstrates a strong correlation between a CEO's educational and career background on the one hand, and the firm's R&D and advertising spending on the other. CEOs with strong technical backgrounds are found in firms with above-average R&D intensities, while advertising expenditures relative to firm sales are high in companies with CEOs that have a marketing background. While some of these effects may be driven by the selection processes which match suitable individuals with particular firms, he also provides evidence of a relationship between a CEO's personal background and the firm's performance. Particular combinations of firm strategy and CEO background are more successful than others. For firms with prospecting behavior, a combination of technical and economic/managerial[2] training is positively associated with success, while exclusive economic or managerial training is detrimental. Somewhat surprisingly, for firms with a defender strategy, exclusive economic or managerial training of the CEO shows an even stronger negative correlation with the firm's performance, while a career background in finance and controlling is helpful. Career backgrounds in legal functions are negatively correlated with performance for prospectors, while career paths in manufacturing are negatively associated with success in defender-type firms. Despite the complexity of these patterns, one surprising result is clear: exclusively economic or managerial training per se is detrimental to performance, both for defender and prospector firms. This is worrisome news for anybody concerned about the societal contribution of business administration as a discipline.[3]

Following Schrader (1995a), this paper starts from the strategic management literature and distinguishes firms with respect to their innovation strategies. A number of hypotheses relate the firm's strategic orientation, characteristics of the top management team, and performance data to each other. The empirical part of the paper makes use of R&D, investment and productivity data from a German innovation survey.[4] The survey also includes detailed data on the

---

[2] The data do not allow Schrader to distinguish clearly between business administration and economics degrees. However, it is plausible to assume that most individuals classified in his sample as having "economic or managerial training" graduated with a Master of Business Administration.

[3] Brockhoff (1998) poses this question for business adminstration in Germany, but expresses doubts that the U.S. results presented by Schrader (1995a) carry over to the German context. See also Brockhoff (1996).

[4] In a previous paper (Harhoff 1997), the same data were used to develop and test an industrial organization model of innovation. The strategic management and the industrial organization views of the world are not necessarily at odds with each other. Industrial organization models typically emphasize that the choice of innovation inputs (e.g. R&D expenditures) precedes output decisions. Thus, the most natural game-theoretical setup involves a first stage in which R&D decisions are made, and a second stage involving output or price decisions. The ex-

firm's innovation objectives. Furthermore, the survey data have been supplemented with information on the academic training of the top managers in the firm. In a number of empirical tests, I explore the relationship between the firm's innovation strategy, managerial education and the firm's performance.

The remainder of the paper consists of five sections. Section 2 provides a more detailed discussion of the theoretical rationale underlying this research. Although the aforementioned conceptual and empirical problems are hard to tackle, a few hypotheses can be tested even with the imperfect data available for this study. The subsequent section describes in more detail the key constructs used here, and discusses briefly the main hypotheses. The data and the variables used to operationalize the three constructs linked here: innovation strategy, managerial education, and firm performance are introduced in section 4. In section 5, several multivariate specifications are presented which allow me to describe the relationship between investment, R&D participation and R&D intensity on the one hand and strategy scores and educational variables on the other. Some of these results merely validate the ex ante assumptions regarding the typology of innovation strategies. Others are interesting in their own right, e.g., the partial correlations between educational background and innovation process inputs such as R&D. This section also contains the empirical assessment of the relationship between performance measures, strategies, and top management educational background. The final section summarizes the results from this exploratory study and tries to reconcile differences between the results obtained here and those from the Schrader study.

2. Basic theoretical aspects

The relationships of interest are depicted in Figure 1 which is adapted from Schrader (1995a, Fig. 14-1).

---

tension pursued here sees strategic choices - the firm's choice of a particular set of innovation objectives - as a very first decision with long-term implications. This view predicts that innovation strategy will be very persistent over time, and that firms will make these choices based on their perception of long-term comparative advantages. These arise in a world of heterogeneity quite easily from the firm's histories and stochastic elements.

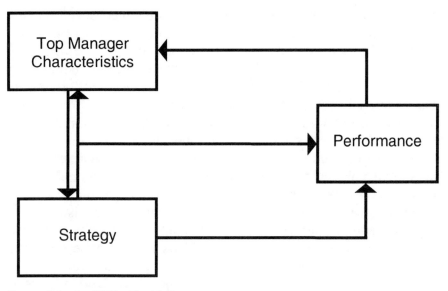

Source: Schrader (1995a, Fig. 14-1)

Figure 1: Linkages between top manager characteristics, strategy, and performance

Strategy and top management characteristics affect each other mutually, and managerial characteristics and the firm's strategic objectives are likely to be correlated for two reasons. First, a firm with a strong focus on, say, technically advanced goods is more likely to attract individuals with comparative advantages to top management jobs in this area. Second, to the extent that a manager has impact on the long-run strategic objectives of a firm, the training and experience a top executive has been exposed to will presumably become important in determining the firm's long-term strategic orientation. A statistically significant partial correlation may be due to either effect - but confirming its sign and significance is already an important result.

Schrader (1995a) suggests that the fit between strategy and top management characteristics will have an impact on firm performance. Moreover, performance will in the long run affect top management in that the firm's owners or supervisory bodies may remove top managers that are persistently less successful than expected. A strong statistical relationship between managerial turnover and the firm's success has indeed been found in many U.S. studies, but also in studies using German data.[5] Another possible mechanism relating performance to management selection are hostile takeovers, which again lead to a drastic change in the composition of the firm's top management.[6] Historically, this mechanism has been more important in the U.S. and the United Kingdom than in Germany or France. Note also that strategy may have an independent effect on performance - some firms (managers) may be better than others in developing strategic responses to changing business conditions.

One could extend the above figure so as to include more explicitly environmental conditions (e.g., the impact of globalization, business cycles, etc.) as a determinant of firm performance, but also as a mitigating variable between top management characteristics, firm strategy, and performance. For example, a technically educated executive may be able to make faster and more precise decisions regarding the choice of a particular technology than somebody not trained in the basics of the respective technical field. During times of considerable technological uncertainty (early in the product life-cycle), a technically trained top manager may therefore be more useful than later on. During a financial crisis, somebody with a background in accounting or controlling may be more successful than an executive with a purely technical education. To keep the figure simple, these relationships have not been depicted in Figure 1.

The logic of the above depiction rests on two major assumptions which deserve brief discussion. The first is that the idiosyncratic characteristics of a top manager will affect his or her decisions and thus the decision-making in the respective organization. This is an utterly plausible idea to build on, and it can be framed as an economic inefficiency of delegation and of information gathering and assessment. If these tasks could be delegated perfectly to a (potentially large) group of analysts - a truly unrealistic scenario - then the respective top manager would be reduced to a mere announcement function. But it is clear from the literature that the assessment and

---

[5]   See for example Jensen and Ruback (1983). For studies on managerial turnover in Germany, see Poensgen (1982), Kaplan (1994), or Schrader and Lüthje (1993). A detailed survey of the turnover literature is presented in Schrader (1995a, ch. 8).

[6]   See - *inter alia* - Jensen (1988) or Franks and Mayer (1995).

framing of information is an integral part of managerial work[7]. Apparently, these activities are extremely difficult to delegate. Principal-agent theory provides a potential explanation for this failure - in the presence of asymmetric information and self-interested managers, the delegating party cannot be fully assured of a decision-making process in line with the owner's or top manager's own interests. Other theoretical foundations which may explain why managerial characteristics matter have been summarized and discussed by Schrader (1995a).[8]

Thus, let us suppose that individuals do matter and that they may have indeed comparative advantages which may be partially contingent on particular environments. If firms (i.e., owners and supervisory bodies) could quickly assess the firm's business situation and the qualities of candidates for a top management position, then they would be able to exchange top managers with more suitable candidates according to the momentary needs of the firm. Again, this is an unrealistic notion. Its implausible flavor points to a second important inefficiency: since managerial turnover is not arbitrarily fast, and since the assessment of a manager's quality takes time, it is unlikely that a firm can shed managers immediately as new business conditions become effective. This principle will be used in this paper to identify the impact of managerial education on performance. More specifically, the German recession of 1993 and the very slow subsequent recovery represent business conditions which were presumably not foreseen when many of the top managers in our sample were selected. This introduces a quasi-experimental effect in the study presented here.

In the long run, one would not necessarily expect performance differentials to last. Over extended periods of time, it should be possible to match heterogeneous problems in the management of an enterprise with the suitably skilled individuals. However, adjusting the

---

[7] See Barnard (1968). The costs of information acquisition and retrieval are also central to Arrow's discussion of alternative modes of economic organization. Cf. Arrow (1974).

[8] Schrader (1995a, ch. 5) discusses and partly integrates models of rational decision-making, leadership, bounded rationality, principal-agent relationships, rituals and symbols in organizations, and models of the firm as a political process. Many of these approaches lend themselves to viewing the personality and characteristics of top managers as important determinants of strategic choices and performance. For reasons of space, I refrain from a discussion here.

human resources of the firm takes time, thus making a short- and medium-term link between educational and other characteristics of managers and performance quite likely. This is particularly true if the time period for which performance is observed is one of a business cycle downturn.

3. Innovation strategy and managerial training - theory and operationalization

3.1 A classification of innovation strategies

This study focuses on the firm's innovation strategy and its relationship to top manager characteristics. Strategy typologies have long been at the center of attention in strategic management research. A special feature of this study is that the empirical classification of firms to strategy types is not based on the firm's actual decisions, e.g., its R&D expenditures or investment behavior, but on a set of survey answers indicating the relevance of particular objectives for the firm's innovation behavior. Separating the strategy classification from actual decision variables is attractive, since it offers the possibility of validating the strategy constructs used here via their correlation with decision variables.[9]

The innovation strategy profiles of firms are characterized along three dimensions in this paper. The first (GLOBAL) seeks to measure the extent to which the firm wants to expand into international markets. The second dimension (PRODUCT) concerns the extent to which the firm focuses on the improvement of product quality, the development of successor products and high market share. Finally, the extent to which the firm concentrates on process improvements (PROCESS) is considered as a third, potentially independent dimension of the firm's innovation objectives.

---

[9] I forego a detailed discussion of "strategy" - the term is used as an abbreviation for firm's long-term objectives and of its focus on particular means of achieving these. This is broadly consistent with Chandler's definition of strategy (cf. Chandler 1962). For a discussion of various other notions of strategy, see Schrader (1995a, ch. 3). A firm's strategic behavior (as discussed in this paper) may be a product of rational choice, trial and error, or pure chance.

These three variables are partly chosen with data availability in mind, but also because they allow the capture of a variety of theoretically appealing typologies in a parsimonious way. For example, industry life-cycle theories of the firm's innovation activities (e.g., Utterback, Abernathy 1975) focus on the potential tradeoffs between product and process innovation. Theories of this type suggest that after an initial phase of intensive product innovation, firms may switch to the pursuit of process improvements in order to achieve economies of scale in production. Schrader (1995a) uses the typology developed by Miles and Snow (1978) very successfully in his study, focussing on prospectors and defenders as the "pure" strategy types.[10] According to this view, firms can assume a prospecting behavior or a more defensive posture. High scores of the GLOBAL measure should be associated with prospecting behavior, while focussing on existing and local markets is likely to reflect a more defensive attitude. Morever, high emphasis on product innovation rather than processes should again be correlated with the more aggressive and dynamic behavior of prospectors, while the converse is true for defenders, who can be expected to focus on efficiency concerns in the narrow sense.

These expectations lend themselves to a simple validation exercise if the strategy classification can be performed without taking recourse to decision variables. In that case, one would expect particularly strong associations between a product-oriented innovation strategy and the firm's R&D intensity, while a focus on PROCESS is likely to cause relatively low R&D intensities and above-average investment efforts. Firms that focus on global rather than local markets (i.e. with high scores in the GLOBAL measure) are expected to have above-average R&D and investment intensities. However, this relationship may depend to some degree on how the firm attempts to cater to foreign markets. In the case of foreign direct investment, the domestic data (which are used here) may not show the expected correlation between global orientation and investment intensity. Conversely, if the firm mainly uses exports of domestically produced goods to achieve its global objectives, a positive relationship in the aforementioned sense should be visible. Since R&D activities are still mainly pursued at firms' headquarters, the resulting pattern for globally oriented firms could be one of above-average R&D inten-

---

[10] The other types are analyzers and reactors.

sity in combination with relatively low investment expenditures (relative to sales).

3.2 Managerial education and training

The introduction of this paper referred to top manager characteristics as being important determinants of decision-making. The characteristics of interest form a relatively large set of demographic and other measures, e.g., parental background, school education, academic training, career background - to name just a few. Schrader (1995a) uses a rather comprehensive set of measures covering both the kind of academic education a CEO received, and the subsequent career orientation. Moreover, in his empirical study he employs the CEO's age and firm tenure as variables in their own right. This paper focuses on a more narrow question: the impact of academic education among the top managers of German enterprises.[11]

The key distinction made in this paper is between individuals who have received academic training in management versus those with scientific or engineering training. As Schrader has argued, these are definitely the two groups of greatest interest for this area of research.[12]

3.3 Major hypotheses

The expectations regarding the relationship between managerial education, the firm's strategy, and actual firm decisions (such as R&D and investment spending) are again guided by the discussion and results put forth in Schrader's work. They can simply be listed as follows:

---

[11] Lack of information regarding career paths is a serious concern, although academic education and career choices are likely to be strongly correlated. Moreover, most of the firms in the sample used here are relatively small - which should increase the correlation between education and career paths. None the less, obtaining a more comprehensive set of data on top manager characteristics, including age and firm tenure, is highly desirable.

[12] Schrader (1995b) focuses on these two groups, since they are the largest ones in the U.S. and in Germany. Moreover, technical and managerial training are profoundly different, having emerged from separate historical developments in each of the two countries.

- a positive correlation between the firm's R&D intensity and academic technical training of its managers, even after including measures of the firm's innovation strategy and other determinants of R&D, such as firm size and age;
- a positive correlation between the firm's investment intensity and academic business administration training of its managers, again after including measures for the firm's innovation strategy and other determinants of its investment policy.

The hypotheses regarding the relationship between performance, education, and firm strategy are more complex, since they implicitly include a contingency or fit assumption. Following Schrader's initial hypotheses (Schrader 1995a), CEOs with technical careers may have a particularly strong positive effect on measures of performance if the firm pursues a strategy that emphasizes technical competencies. For firms with defender strategies, one would then expect CEOs with backgrounds in controlling and finance to contribute positively to firm performance. However, Schrader's results do not show this strong fit pattern, although they generally support the notion that the CEO characteristics of successful defender and prospector firms differ significantly. As discussed before, the most stunning common result is that exclusive economic or managerial training is negatively associated with performance, both for defender and prospector firms.

Rather than identifying subgroups of firms that come close to the ideal of pure prospecting or defensive behavior, the empirical test used in this paper uses the three strategy variables (PRODUCT, PROCESS, GLOBAL) to characterize a firm's strategic orientation along three dimensions. The "fit hypothesis" may in this case be operationalized using interaction terms between managerial education variables and strategy variables. Significant explanatory contributions of these interaction variables, e.g., in an equation explaining the observed productivity change of the firm, are then evidence supporting the fit hypothesis. More specifically, we would expect the interaction between technical training and PRODUCT focus on the one hand, and between managerial training and PROCESS on the other hand, to have positive coefficients. If the interaction terms do not reach statistical significance, but the education variables by themselves carry a significant coefficient, then the respective group of managers has a comparative advantage irrespective of the firm's pursued innovation strategy.

4. Empirical aspects

4.1 Data source

The data used in this chapter originate from the *Mannheim Innovation Panel* (MIP). The data collection is based on mailed survey questionnaires. The survey was commissioned by the German Ministry of Education, Science, Research and Technology and the European Union. It has been carried out annually since 1993 by the Zentrum für Europäische Wirtschaftsforschung (ZEW) in Mannheim. The data cover both East and West German enterprises, though this paper restricts attention to an analysis of West German firms. Moreover, the data encompass both innovating and non-innovating enterprises. A central concern of the MIP survey is to provide a broad qualitative and quantitative description of innovation processes in German manufacturing firms. Towards that objective, it contains questions regarding the relevance of information sources, innovation impediments, and innovation objectives.

The sample used here contains 1992 data on 1,245 West German firms. The median number of employees in this sample is 255, and the median sales volume is DM 57 million. Sixty-nine percent of the firms conduct R&D on a regular basis, and the average R&D intensity of these R&D performers is 1.64 percent. The vast majority of the sample firms is in the legal form of corporations with limited liability. The performance measures (sales growth and labor productivity growth) are computed from 1992 and 1996 data (collected in the 1993 and 1997 surveys, respectively). Since not all firms responding to the 1993 survey participated in the 1997 survey, performance measures are available only for 385 enterprises.[13]

---

[13] Ideally, one should test for selection problems arising from this attrition process. As a first indication that problems are unlikely to emerge, the firm size and age distributions do not display significant differences between firms that were represented in both years and firms that only participated in 1993.

4.2 Innovation strategy

While it would in principle be feasible to follow Schrader (1995a) in classifying strategy types on the basis of R&D and other expenditure variables, this paper follows a different approach which uses qualitative survey responses directly. Table 1 summarizes the computation of the strategy variables. The MIP survey contains a question in which participants are asked to rate the relevance of various innovation objectives (listed in Table 1) on Likert-type scales. This information is aggregated to the three scale variables discussed above.

Table 1 does not include the full set of innovation objectives. I excluded the variables "reduction of environmental problems in production", "improvement of work conditions", and "development of environmentally friendly products", since they cannot be assigned easily to the three theoretical constructs. Also excluded were variables measuring the relevance of local markets in West and East Germany. Including these actually reduced the average inter-item correlation in all of the three scale variables drastically, since these two items display very little variation across firms. The Cronbach Alpha values for the three scales as summarized in Table 1 are 0.61 (PRODUCT), 0.83 (PROCESS), and 0.84 (GLOBAL). Thus, the latter two are reasonable, while the first scale could profit from improvements in further work.[14] The three scale variables enter the regressions below in standardized form.

---

[14] Initially, the strategy variables were computed via factor-analytic methods. The overall results are quite similar to those presented here in the multivariate analysis.

Variable	*Overall relevance*	Scale 1 PRODUCT	Scale 2 PROCESS	Scale 3 GLOBAL
Develop successor products	72.4%	1	0	0
Increase or maintain market share	91.5%	1	0	0
Extend product range within existing focus	69.7%	1	0	0
Extend product range beyond existing focus	23.3%	1	0	0
Improve product quality	84.4%	1	0	0
Access new markets in Eastern Europe	26.7%	0	0	1
Access new markets within the EU	53.4%	0	0	1
Access new markets in Japan	14.9%	0	0	1
Access new markets in the US/Canada	28.0%	0	0	1
Access new markets in other countries	25.5%	0	0	1
Increase flexibility of production	65.6%	0	1	0
Reduce costs/lower wage bill	74.1%	0	1	0
Reduce costs/lower materials consumption	59.4%	0	1	0
Reduce costs/lower energy consumption	43.8%	0	1	0
Reduce costs/lower production setup costs	53.1%	0	1	0
Reduce costs/lower reject rate	66.9%	0	1	0

Note: *Overall relevance* is the share of firms which rated the respective objective as important or highly important (see text for further explanations). The figures in the three subsequent columns reflect the weight of individual innovation objectives in the computation of the scale variables.

Table 1: Computation of innovation strategy scales

4.3 Managerial education

While CEOs in U.S. corporations have a singularly influential position and are, thus, ideally suited for a study like the one put forth by Schrader (1995a), it is less clear which individuals should be identified as the relevant top managers in German companies. In German public stock companies, the members of the board usually govern the firm's affairs in a more egalitarian way than in U.S. companies. However, the data used here cover mostly small and medium-sized firms, where this issue will be less problematic. Since most firms are in the legal form of a limited liability corporation, the top managerial functions are well defined in legal terms.[15] The names of the respective individuals must be registered with the commercial registrar. The coding of academic degree information relies on the assumption that the top managers will - on this occasion - indicate the full title of the degrees they have received. Given that academic degrees presumably play a much larger role in Germany than in other countries, this assumption is not a critical one. The commercial registrar thus contains this information which has been matched with the innovation survey data for the purpose of this study. If the data include the names of several top managers (as is typically the case), all of the available information is used to compute the share of individuals with a particular degree.

The variables generated from these data are ENGINEERING DEGREE, BUSINESS ADMINISTRATION DEGREE, OTHER TECHNICAL/SCIENTIFIC DEGREE and OTHER ACADEMIC DEGREE. In all cases, only academic degrees were considered, and the reference group are individuals without any such degree. As pointed out before, the focus in this study is on the first two groups. They are also the largest ones: the engineering degree variable has a sample mean value of 0.40, the business administration variable a mean of 0.21. The reference group (no academic degrees) is the second largest with a mean of 0.33. This distribution should not be surprising. After all, many of the German *Mittelstand* firms are known to have strong technical traditions.

---

[15] These functions are the *Geschäftsführer* and the *Prokurist*.

A few comments on the limitations of the data are in order here. Combined degrees were hard to detect given that the data typically reveal the terminal degree of the manager. Furthermore, we only have available the degree information and no additional data on years of schooling, types of schools attended prior to receiving the final degree, etc. While this limits the usefulness of the data at this point, future studies will also have data on the size of the management team, the manager's age and other information available. This restriction is currently mirrored in the title of the paper, declaring it an exploratory study. The study should be seen only as a first test using German data. As will become clear in the next section, the results are quite strong despite the present data limitations.

5. Empirical analysis

5.1 Investment in plant, property and equipment

Table 2 contains the results from simple OLS regressions, using the logarithm of the investment intensity as the dependent variable.[16] The performance of these equations is not particularly good, even if one takes into account that cross-sectional regressions of investment produce relatively low R^2 measures. The overall measures of fit are not dismally weak, but the industry dummy variables are mostly responsible for the R^2 values. However, even after controlling for industry effects (using 21 industry groups), there are still indications that the strategy classification works as intended, albeit in a statistically weak form. Comparatively high values of the PROCESS scale variable are indeed associated with high investment intensities ($p<0.1$). Firms focusing on global markets, however, have a lower investment intensity. Given that some of the globally oriented firms may engage in foreign direct investment, rather than domestic production and export sales, it is clear that we need a more refined model which includes export activity. Since the simultaneity problems are by no means trivial when an export measure is included among the regressors, this has not been attempted yet.

---

[16] Investment is defined as additions to plant, property and equipment.

Independent Variable	(1)	(2)
PRODUCT	0.058	0.060
PROCESS	0.050*	0.050*
GLOBAL	-0.053*	-0.052*
ENGINEERING DEGREE	-	0.016
BUSINESS ADMINISTRATION DEGREE	-	-0.088
OTHER TECHNICAL/SCIENTIFIC DEGREE	-	-0.147
OTHER ACADEMIC DEGREE	-	-0.038
ln(employees)	-0.010	-0.006
ln(firm age)	0.011	0.011
Joint Test for Industry Dummy Variables (p value)	$p<0.001$	$p<0.001$
Joint Test for Innovation Strategy Variables (p value)	$p=0.118$	$p=0.101$
Joint Test for Degree Variables (p value)		$p=0.685$
Adjusted R^2	0.076	0.074
Joint Test of Slope Parameters (p value)	$p<0.001$	$p<0.001$

Note: 1217 observations. All variables computed from the 1993 cross-section of the *Mannheim Innovation Panel*. Only West German firms were included in the sample. The degree variables measure the proportion of top managers with the respective degree. The reference group are managers without any academic education.

Significance levels (two-tailed tests): * $p=0.1$, ** $p=0.05$, *** $p=0.01$,

Table 2: Regression analysis of investment behavior
Dependent variable: ln (investment/sales)

5.2 Continuous R&D activities and R&D intensity

A second test of the innovation strategy variables is contained in Table 3 and Table 4. Exploring the relationship between the firm's R&D behavior and the strategy variables yields more satisfactory results than before. There are actually two parts to this test. In Table 3, the probability that the firm undertakes R&D on a regular basis (continuously) is modelled as a function of manager education and the firm's innovation strategy. In Table 4, the R&D intensity of R&D performing firms (again in logarithms) is related to the same variables. I first turn to the results from the probit specification in Table 3, where the coefficients have been recomputed to reflect marginal effects. Here, all three strategy variables are highly significant and carry signs which are consistent with the theoretical arguments developed before. Firms with a strong focus on global markets and firms that put emphasis on developing new and improved products are more likely to undertake R&D than firms which focus on the optimization of production processes. Since the strategy variables are standardized, the coefficients can be interpreted quite easily. For example, a change in the PRODUCT score by one standard deviation will increase the probability of continuous R&D activities by 10.9 percent. Interestingly, the R&D participation is almost entirely driven by the strategy variables; there is barely any influence from the degree variables. Only in the case of technically trained top managers can we detect a marginally significant greater likelihood of pursuing continuous R&D activities. The overall predictive power of the strategy variables is quite high, as the pseudo-R^2 values indicate.

Independent Variable	(1)	(2)
PRODUCT	0.133***	0.133***
PROCESS	-0.085**	-0.084**
GLOBAL	0.109***	0.108***
ENGINEERING DEGREE	-	0.053*
BUSINESS ADMINISTRATION DEGREE	-	-0.005
OTHER TECHNICAL/SCIENTIFIC DEGREE	-	-0.080
OTHER ACADEMIC DEGREE	-	0.031
ln(employees)	0.112***	0.111***
ln(firm age)	-0.001	-0.002
Joint Test for Industry Dummy Variables (p value)	p<0.001	p<0.001
Joint Test for Innovation Strategy Variables (p value)	p<0.001	p<0.001
Joint Test for Degree Variables (p value)	-	p=0.381
Pseudo R^2	0.265	0.268
Joint Test of Slope Parameters	p<0.001	p<0.001

Note: 1245 observations. The coefficients represent marginal effects. All variables computed from the 1993 cross-section of the *Mannheim Innovation Panel*. Only West German firms were included in the sample. The degree variables measure the share of top managers with the respective degree. The reference group are managers without any academic education. Significance levels (two-tailed tests): * p=0.1, ** p=0.05, *** p=0.01

Table 3: Probit analysis of continuous R&D activities
Dependent variable: firm continuously undertakes R&D (0/1)

The picture is slightly different in the case of the firms' R&D intensities (Table 4). We observe again that the strategy variables are both large and highly significant. But in the specification in column (2), the coefficient of the engineering degree variable carries a positive sign and is highly significant ($p<0.01$). It is noteworthy that a coefficient of similar size materializes for the group of firms with top managers that have other technical or scientific degrees. However, the associated standard error is quite large. It is possible that this group is highly heterogeneous, and that the within-group heterogeneity leads to imprecise measurement.

Taking the results from Table 3 and Table 4 together, the theoretically expected signs for the coefficients of the strategy variables do actually appear. It is astonishing that the technical degree variable is significant here, since the innovation strategy variables are defined as continuous measures, not just grouping variables.

Independent Variable	(1)	(2)
PRODUCT	0.279***	0.267***
PROCESS	-0.223*	-0.212***
GLOBAL	0.269***	0.274***
ENGINEERING DEGREE	-	0.324***
BUSINESS ADMINISTRATION DEGREE	-	-0.099
OTHER TECHNICAL/SCIENTIFIC DEGREE	-	0.347*
OTHER ACADEMIC DEGREE	-	-0.128
ln(employees)	-0.044*	-0.048*
ln(firm age)	-0.062	-0.063
Joint Test for Industry Dummy Variables (p value)	p<0.001	p<0.001
Joint Test for Innovation Strategy Variables (p value)	p<0.001	p<0.001
Joint Test for Degree Variables (p value)	-	p=0.012
Adjusted R^2	0.201	0.210
Joint Test of Slope Parameters	p<0.001	p<0.001

Note: 898 observations with non-zero R&D expenditures. All variables computed from the 1993 cross-section of the *Mannheim Innovation Panel*. Only West German firms were included in the sample. The degree variables measure the share of top managers with the respective degree. The reference group are managers without any academic education.
Significance levels (two-tailed tests): * p=0.1, ** p=0.05, *** p=0.01

Table 4: Regression analysis of R&D intensity
Dependent variable: ln (R&D expenditures/sales)

5.3 Performance, strategy, and managerial education

The panel nature of the data is finally exploited in this subsection. The analysis of revenue growth between 1993 and 1996 and of labor productivity growth across the same period yields an unexpected, yet statistically robust result.[17] For each of the two dependent variables, two empirical specifications are used. The first one contains the full set of variables without strategy-education interaction terms. The second column performs a test of the "fit hypothesis" according to which the specific experiences and talents of managers will be particularly effective if the firm pursues a complementary strategy. The results in Table 5 do not support the fit hypothesis. But they do show clear performance advantages for firms that were mostly led by top managers with academic training in business administration. Consider column (1) in which revenue growth is regressed on innovation strategy variables, indicators of academic training, and control variables. The only variables that carry significant coefficients are firm age and the educational variable for business administration training. Including the interaction terms between the strategy and degree variables in column (2) does not improve the statistical picture. This equation would tentatively suggest that revenue growth was stronger in firms led by academically trained business administrators. The coefficient of 0.16 translates into an annual growth difference of about 4 percent.

Columns (3) and (4) complement this information with labor productivity *growth* equations. The business administration variable is again significant ($p < 0.05$), and the differential in growth rates is virtually the same as in the revenue growth specification. In other words: the same firms that expanded at an above-average rate were able to restrict labor input to 1993 levels. Virtually all of the revenue growth was transferred into labor productivity growth. Large firms were apparently more successful in reducing their workforce than smaller firms. These results do not change once a full set of 21 in-

---

[17] Labor productivity is defined here as sales per employee. The growth rates are computed as logarithmic differences. Therefore, labor productivity growth can be decomposed into two additive components: revenue growth and employment growth. Adding a third equation that models firm growth in terms of employment would therefore be redundant.

dustry dummy variables is included. The test on joint significance of the industry variables yields very small F statistics, with significance levels always beyond 0.30.

Independent Variable	Revenue Growth		Labor Productivity Growth	
	(1)	(2)	(1)	(2)
GLOBAL	-0.036	-0.037	0.024	0.024
PROCESS	-0.058	-0.047	-0.027	-0.015
PRODUCT	0.022	0.011	-0.054	-0.052
ENGINEERING DEGREE	0.097	0.100	0.046	0.046
BUSINESS ADMINISTRATION DEGREE	0.161*	0.158*	0.133**	0.131**
OTHER TECHNICAL/ SCIENTIFIC DEGREE	-0.071	-0.069	-0.053	-0.054
OTHER ACADEMIC DEGREE	-0.018	-0.014	-0.144	-0.15
ENGINEERING DEGREE*PRODUCT	-	-0.041	-	-0.011
BUSINESS ADMINISTRATION DEGREE*PROCESS	-	0.026	-	-0.005
ln(employees)	-0.025	-0.025	0.031**	0.031**
ln(firm age)	-0.070**	-0.070**	-0.029	-0.029
Joint Test for Innovation Strategy Variables (p value)	p=0.41	p=0.42	p=0.36	p=0.59
Joint Test for Degree Variables (p value)	p=0.30	p=0.31	p=0.22	p=0.24
Joint Test for Interaction Variables (p value)	-	p=0.90	-	p=0.99
Adjusted R^2	0.026	0.021	0.049	0.049
Joint Test of Slope Parameters (p value)	p=0.025	p=0.058	p=0.015	p=0.041

Note: 385 observations. All variables computed from the 1993 and 1997 cross-sections of the *Mannheim Innovation Panel*. Only West German firms were included in the sample. The degree variables measure the share of top managers with the respective degree. The reference group are managers without any academic education.

Significance levels (two-tailed tests): * p=0.1, ** p=0.05, *** p=0.01

Table 5: Regression analysis of revenue and labor productivity growth 1992-1996

Taken together, these results suggest that ceteris paribus firms with top managers who had received academic training in business administration were able to expand their output while keeping their labor roughly constant. Firms that were led by managers with other types of academic training were apparently not able to replicate this performance.[18]

6. Discussion and conclusions

The results summarized in this paper point to two intriguing empirical regularities. Firstly, technical training of top managers is positively correlated with the firm's R&D intensity, even after controlling for the firm's strategic orientation as measured by the three strategy variables. This result is not too surprising, and it should not be interpreted in any causal way. The strategy variables may simply fail to capture fully the firm's focus on technological comparative advantage, and technically trained managers may be attracted to firms that follow such a path. But the strong correlation is interesting in its own right, and it documents that key decisions in technology-oriented firms tend to be undertaken predominantly by technical experts. This finding corresponds nicely to the empirical results presented by Schrader (1995a).

The second result concerns the link between the performance of firms and the academic education of its top managers. The exploratory regression results documented here show a clear advantage for firms which were led by personnel with university-level training in management. These firms tend to show significantly higher sales growth from 1993 to 1996. But their labor productivity also grew much more strongly than that of firms which were directed by technically trained individuals. This result is apparently at odds with Schrader's (1995a) conclusion.

---

[18] The low R^2 values in these regressions are typical for panel data. Including a full set of 21 industry dummy variables does not lead to a change in the results. The industry dummy variables are jointly insignificant in all the four specifications in Table 5.

How can the differences be reconciled? There are two good reasons why both results may be correct.[19] First, they refer to two different countries with substantial differences in academic managerial training. As Schrader (1995b) and Brockhoff (1998) point out, the management training in Germany is typically broader and more general than MBA programs in the U.S. The German study programs typically also put more emphasis on a thorough scientific foundation of managerial principles, while the U.S. programs tend to be designed with immediate application of the conveyed skills in mind.

Independent of this interpretation, there may be a second one that could possibly explain the differences even if the educational systems in Germany and the U.S. were to produce the same type of business administration graduate. Particular skills and career experiences may endow an executive with comparative advantages which can only be realized under certain circumstances. In periods of strong growth, technical decisions may be of more relevance for the profitability and growth of a firm. In financially tough times managerial skills in cost accounting, controlling, and marketing may be more productive. The German data used here cover the years from 1992 to 1996, a period that includes a deep recession in Germany combined with considerable labor shedding. During this period, many German firms had to adjust to increasing competition and cost pressure. At least in the short run, engineers may be less suited to manage this adjustment process than business-trained executives.

The overall results of this exploratory study, taken together with the much more comprehensive study conducted by Stephan Schrader are somewhat paradoxical. When Stephan Schrader (who received his academic training in *Betriebswirtschaftslehre*) embarked on his project to link firm performance and strategy to the educational and career background of CEOs in the United States, he did not anticipate that he would find a positive correlation between a technical background and performance. When this author (whose first training was in mechanical engineering) started the analysis summarized in this paper, he anticipated (and personally may have liked) to find a

---

[19] Alternatively, one may argue that either one or both studies get it wrong. Both or one of the two results may be spurious, e.g., due to selection issues or other data or methodological problems. In this respect, one ought to have more confidence in the U.S. results presented by Schrader, since the data he used are more comprehensive and less selective.

similar result, i.e., that engineering-dominated firms had better performance. Quite the contrary materialized. It seems that the analysis of this topic is apt to defeat expectations - which should make further analysis of the nexus between executive characteristics, firm strategy, and performance even more appealing. After all, a problem worthy of attack bites back.

Acknowledgements

This paper was prepared while the author was Research Professor at the Social Science Research Center Berlin (WZB). I would like to thank the Center for its hospitality and research support. Georg Licht provided many helpful suggestions. The topic of the paper and the content owes much to Stephan Schrader's insightful and thought-provoking research on executive demographics and firm performance. Constructive comments made by the participants of the conference on the Dynamics of Innovation Processes, held in Hamburg on July 16/17, 1998 helped to improve the paper considerably.

Part II: Dynamics in project management

Promotors and champions in innovations – development of a research paradigm

Jürgen Hauschildt

Promotors and champions in innovations - development of a research paradigm

Jürgen Hauschildt

1. Phases of development

1.1 Phase 1: The discovery of the champion

Josef Schumpeter can be credited with being the first person to draw attention to the central role played by the *entrepreneur* in innovation processes, in a book published in 1912 (Theory of Economic Development, Leipzig). This entrepreneur creates new combinations on a discontinuous basis, in totally new forms, in an act of creative destruction. He brings forth new products, introduces new production methods, opens up new markets, conquers new sources of supply or re-organizes (Schumpeter 1931, p. 100). The "dynamic entrepreneur" was thus characterized and described. This was apparently sufficient to incorporate him into the economic models. To understand him as a real person or even to analyze him in further detail seemed superfluous.

That changed when, in 1963, Schon introduced a new term for this creative individual: the *"champion"*. The term which has dominated discussion to date was thereby established for the Anglo-Saxon countries. In contrast, in the German-speaking countries this term was not accepted because of a slightly negative connotation.

Some ten years had elapsed when, almost simultaneously in three different places in the world, mutually independent studies confirmed that the activity of committed and enthusiastic individuals plays a decisive role in promoting the creation of innovations.

- In *Germany*[1] in 1973, Witte investigated the first purchases or leasing of computers. In his survey he proved that the existence of "promotors" – as he termed the champions – led to significantly higher levels of innovation and of activity than found for innovation processes in which such individuals were absent ("Columbus" Project).

- In the *USA* in 1974, Chakrabarti discovered that product champions could be found chiefly in successful cases in the further development of NASA innovations ("NASA Study").

- In *England* in 1974, Rothwell and his team, conducting research into innovations in chemical processes and scientific instruments, found that the human factor was a key determinant for the success of innovation (SAPPHO Project).

The breakthrough was thus achieved: the importance of the human factor was established beyond doubt. Many research projects, particularly the study by Howell/Higgins, have confirmed over and over again that identifying the champions or promotors in the innovation processes is not a great problem. They normally stand out clearly because of their original contributions and/or because they make quite deliberate use of their power to further the innovation process. Consequently, it is easy to identify the active individuals in the innovation processes. *Champions are no longer merely literary figures, but empirically observable individuals who can be described using suitable quantification conventions and who are clearly successful.*

1.2 Phase 2: Confusion

These three studies were not the only ones, but they were the ones which dealt most clearly with the human side of innovation. Other studies published in the early 1970s by Rogers/Shoemaker (1971), Langrish et al. (1972), Globe et al. (1973), and Havelock (1973) should also be mentioned. The common feature of all these studies is that they identified not just one single champion in an innovation process, but several outstanding individuals who were present si-

---

[1] For the development of German empirical research on promotors in the last 25 years see Hauschildt, Gemünden 1998.

multaneously in innovation processes. Different terms were sought to distinguish these committed individuals from one another. The initial consequence was that a confusing *variety of terms* appeared in the literature, often colored by the language used in normal practice. The following is a small selection:

Inventor, initiator, stimulator, legitimizer, decision maker, executor, catalyst, solution giver, process helper, resource linker, technical innovator, product champion, business innovator, chief executive, technology promotor, power promotor.

This flood of terms for those people who actively promote innovation processes has by no means died to a trickle since then. More terms can also be found in new publications of the '90s:

Political coordinator, information coordinator, resource coordinator, market coordinator, management champion, decider, planner, user, doer, expert, person affected, process promotor, relationship promotor.

Roberts and Fusfeld (1981) did not just describe this variety of terms, they also had them caricatured in a particularly impressive manner.

But what was the outcome?

On the *negative side*, complaints included confusion, lack of clarity, redundancy of terms and encouragement of different schools of thought. Researchers are not exempt from the ambition to establish "their" terms and to provide evidence that the distinctions they have selected are particularly useful either for further research or even for direct application.

On the *positive side*, variety became apparent. The fact that a different number of individuals with different functions would be found in different innovation processes was established as a certainty. As a result, the question of the cause and effect of these differences could be raised.

1.3 Phase 3: Order

When Alok Chakrabarti joined us at the Institute for Research in Innovation Management in Kiel in 1987, we both stood amazed before this bewildering variety. We saw it as our first task to establish order so that further research could follow a systematic concept. This concept had to take into account the two functions of an organization: first, to efficiently regulate the *work* to be done, second, to effectively regulate the *power* relationships between the incumbents. We took the terms for the people engaged in innovation processes to express certain *activities* performed by them during the process, or certain *power bases* from which they derived their influence on these innovation processes. This produced the following twofold distinction by contributions and power bases:

Activities and roles in innovation management	
Activities	**Roles in innovation management**
1. Initiation of innovative process	Initiator, catalyst, stimulator
2. Development of a solution	Solution finder, solution giver, idea generator, information source
3. Process management	Process helper, connector, resource linker, idea facilitator, orchestrator
4. Decision making	Decision maker, legitimizer
5. Implementation	Realizer, executor

Source: Hauschildt/Chakrabarti (1988, p. 383)

Power bases and roles in innovation management	
Power bases	**Roles in innovation management**
1. Knowledge specialty	Technology promotor, technical innovator, technologist, inventor
2. Hierarchical potential	Power promotor, chief executive, executive champion
3. Control of resources	Business innovator, investor, entrepreneur, sponsor
4. Organizational know-how and communication potential	Process promotor, product champion, project champion
5. Network know-how and potential for interaction	Relationship promotor

Source: Chakrabarti/Hauschildt 1989, pp. 165-166; Gemünden/Walter 1995, pp. 973 ff.

Table 1: Roles in innovation management

1.4 Phase 4: Explanation of success

(1) The first studies on the champions in the Anglo-Saxon countries differed from the German promotor studies in one very significant point: in the publications of Rothwell et al. and Chakrabarti, the *paired comparison approach* was used. Successful companies or projects were compared with companies and projects which produced little or no success. Attempts were made to identify the characteristics which distinguished successful cases from unsuccessful ones. This approach was undoubtedly focused, drawing on the basis of everyday experience and academic topologies, but was not driven by concise theory and did not test a theory in the strict sense.

This is the most prominent difference from the German research conducted under Witte (Witte 1977, pp. 47 ff.). He first of all developed an original *theoretical concept*, which explains why the presence of promotors improves the success of the innovation process. Witte worked with the hypothetical construct of *barriers*: resistance to the innovation due to the barrier of *ignorance* and due to the barrier of *unwillingness*. Promotors commit enthusiastically to the innovation and help to overcome these barriers. The promotor model contains three core theorems:

1. Each type of resistance has to be overcome by a specific type of energy. The barrier of unwillingness is overcome by *hierarchical potential*, the barrier of ignorance is overcome by the use of *specific knowledge in a certain technical field* (correspondency theorem).
2. These types of energy are provided by different people. The *power promotor* ("Machtpromotor") contributes resources and hierarchical potential and the *technology promotor* ("Fachpromotor") contributes specific technical knowledge to the innovation process (theorem of division of labor).
3. The innovation process is successful when the power promotor and technology promotor form a *coalition* and are *well coordinated*, i.e. when they really co-operate (theorem of interaction).

The promotor model is thus based on the specific use of *power bases*. In addition, however, close cooperation between the promotors is also important. Witte chose the term "tandem structure" (or "dyad") for this, in the sense of two horses harnessed to a carriage in tandem.

Using a sample of 233 initial acquisitions (by purchase or lease) of computers, the empirical test showed that not only were much more innovative solutions found, but that the work also proceeded much faster and with greater diligence in those cases where such a tandem structure was present (Witte 1973).

(2) It is undoubtedly true to say that one significant contribution made by promotors lies in overcoming resistance to an innovation. However, this area is also a target of criticism of Witte's concept: the promotors do more than just cope with conflicts. This is particularly true when the opposition, overall, has a loyal attitude, as the findings of Markham et al. (1991) prove. The original promotor model was in essence a conflict-handling model. However, this view distracts from the informative and creative aspects of innovations. After all, innovations are particularly characterized by the fact that information is completely newly generated and/or combined in them. Furthermore, innovations are processes of problem definition, goal formation, generation and identification of new combinations. When Witte's model was developed, these *cognitive tasks* were given less consideration than the *conflict-handling functions* of the promotors. The cognitive tasks could probably supply a different theoretical base for the interaction among the promotors.

In 1992, the research by Ancona/Caldwell went into this *interaction of cognitive and conflict-handling activities by the champions* in more detail. The Ancona/Caldwell study determines four characteristic areas of activity by factor analysis:

- "Ambassadorial activities": Formation of goals and blocking of opposition, above all conflict-handling activities;
- "Task coordinator activities": Coordination, negotiation and interface management, also basically conflict handling;
- "Scouting activities": Obtaining information, building expertise, seeking solutions, clearly cognitive activities;
- "Guard activities": Prevention of an undesirable leak of ideas and information, activities which are not covered by our concept.

This seems to us to provide sufficient evidence of the cognitive contributions of the promotors. The comprehensive model explaining the human influence on the innovation process must definitely combine cognitive and conflict-handling activities.

1.5 Phase 5: Systematic differentiation of the division of labor in contingency models

(1) The variety which emerged in the wake of the Witte, Rothwell and Chakrabarti studies raised two questions: not only: "What effect does such variety have?", but also "What determines it?". The traditional *contingency view* of organizational theory could thus also be applied to innovation management. The next question was: *How do external circumstances affect the number of promotors and the way in which they approach the division of labor?*

(2) Witte's findings had already indicated that the division of labor between the technology promotor and the power promotor was clearly a phenomenon of *corporate size*. Rothwell and his research team proved that the *industry* is a determinant of division of labor. The *degree of innovativeness* and the *degree of diffusion* of the innovative products or processes also influence the division of labor. With the increasing diffusion of the innovation in an economy, the importance of the technology promotor declines. Maidique arrived at similar results as early as 1980 (see table 2).

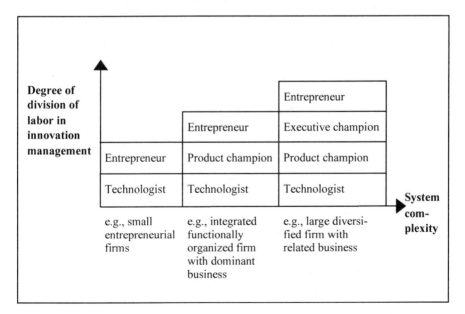

Table 2: System complexity and division of labor in innovation management
(Maidique 1980)

If we put these findings together, we find *two influences super-imposed* which are important for the division of labor in innovation management: *system complexity and problem complexity.* The resulting *overall complexity* has to be compared with the capability of the active individuals. If there is a considerable discrepancy between the complexity of the innovation and personal innovative capability, the basic model of division of labor has to be modified.

Maidique dealt with the type of instance where more extensive division of labor becomes an inevitable result of *high system complexity* as early as 1980. In the simplest case, in a small, entrepreneurial company, a *two-center constellation* of technology promotor and power promotor is found. According to Maidique, a *three-center constellation* is typical in medium-sized companies with a functional structure which are still limited to one product line. A *four-center constellation* is to be found in very large, diversified companies.

(3) The many and varied completed research projects which have been analyzed by Chakrabarti/Hauschildt (1988) contained many references to three-center constellations. To find an explanation we applied the complexity concept, based our work on Witte's concept and identified a third species of promotor: the *process promotor.* Process promotors are needed when innovations affect a very large number of individuals personally in relatively large institutions, and trigger conflicts. Like the other promotors, process promotors rely on specific power bases: on system know-how, organizational and planning power, and on interactive skills. They, too, overcome characteristic forms of resistance: those of an established organization whose aim is to execute routine procedures as efficiently as possible and which rejects innovations as a disruption of its smooth running. Process promotors do not have the formal authority of the power promotor or the expertise of the technology promotor. They rely on leadership qualities and influencing tactics, and like the other two promotors they are characterized by the fact that they undertake risk and are prepared to sink or swim with the innovation. The study by Howell/Higgins demonstrated this side of the champion in particular. The Gesche Keim study confirms this view for the German-speaking countries: the successful "interactive project managers" are particularly characterized by a high level of interactive skills, cooperative leadership, above-average problem-solving capabilities and constructive creativity (1997, p. 161).

Building on Witte's concept of the "tandem structure" we call the team of three the "*troika*" *of power, process and technology promotor*. In his study of 133 innovation projects in the mechanical engineering industry, Kirchmann (1994) proved that this troika structure achieves better technical results, but above all better economic results, than any other structure. Lechler (1997) confirmed these findings in his research into 448 projects, with an interesting addition: he was able to confirm our assumption that the probability of the occurrence of a process promotor and his positive influence on a project's outcome increases with problem complexity.

Figure 1 summarizes the conflict handling and cognitive activities in the division of labor between the promotors in the troika structure.

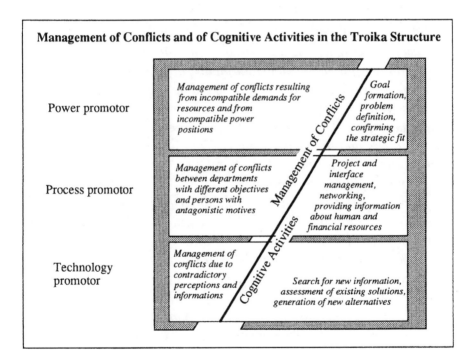

Figure 1: Management of conflicts and of cognitive activities in the troika structure

(4) Finally, Gemünden and Walter (1998) indicate a further modification of the troika concept: they point out that more and more innovations require *cooperation with external partners* in the value chain, i.e. with customers or suppliers. There are barriers in the way of this cooperation, too. Just as promotors are needed to eliminate in-house barriers, they are also called upon to overcome extramural barriers of interaction. The barrier concept is developed similarly to Witte's theoretical approach. In place of the process promotor, who overcomes in-house barriers only, Gemünden and Walter's concept includes the "*relationship promotor*". In a study of 94 technology transfer projects, they proved that processes are more successful if a person is present who deliberately establishes and maintains relationships with the partners. This proves their efficiency, at least for the function of the relationship promotor, although no proof of the division of labor is available as yet, such as that supplied for the troika of power, process and technology promotor.

In summary: research initially proved that an active, committed champion was the most important factor for success in the management of innovations. However, at the same time a variety of other persons were observed who were also striving to make the innovation successful. These individuals could be distinguished by their contributions and their power bases. The successful impact of these individuals is due to their skills in dealing with conflicts constructively and handling information creatively. Finally, we know that the extent and type of the division of labor among these individuals is determined by the complexity of the innovation problem to be solved and by the complexity of the organization concerned. Accordingly, the troika structure consisting of power promotor, process promotor and technology promotor in particular is the most successful structure for the in-house management of typical innovation projects. It is possible that the increase in extramural innovation activity will shift the role of the process promotor more towards that of a relationship promotor.

2. Possible routes for future research

As we ask ourselves "How is the research process likely to move forward?", we are entering the realms of science fiction. One can see four routes in future, three of which are theoretical, addressing the problem of *explanation*. The fourth route is directed at the *application* of the promotor or champion concept in practice. Let us first of all go down the theoretical routes:

2.1 Route 1: Further details on the promotor model

Any serious academic will have no great difficulty in spontaneously reeling off a list of questions which the current body of research cannot answer, or cannot answer satisfactorily. It follows logically that a whole generation of academics can be occupied with identifying and proving further details of the promotor model. We see the following as the most pressing questions:

- Which *key events* stimulate individuals to act as promotors of innovations?

- *How do promotors come together*? The first phase of innovation processes is normally lost in mystic obscurity. Yet it is in this phase that the process by which promotors come together takes place, a process which can obviously only be described in social and psychological categories.

- And once these promotors do actually encounter one another - how is the *personal fit* determined and secured? A good fit is essential for the subsequent innovation project to come to a successful conclusion with all its difficulties, and to get it completed in the face of all resistance. Such teams need considerable group cohesion in order to withstand all the pressures from outside.

- We know very little about the conditions under which promotor structures are *dissolved*. Even if a promising promotor team comes on the scene at the beginning of a process, it is by no means certain that it will see the process through and complete it successfully. On what reefs might the tandem or troika founder?

- The interaction of the promotors and champions is couched in somewhat mysterious terms as "*good cooperation*". What does that mean specifically? We know very little about whether and how the individual promotors have to take a hand during the innovation process. Do they play changing roles? Do they always appear as a team? What contribution is absolutely essential for which key occurrences?

– Does the promotor model apply *regardless of time and space*? Organizations have changed in the last 20 years: they have become more open, more tolerant of conflict, more process-related, more project-orientated, more targeted, more risk-aware, more informative, and more cooperative. New forms of organization have developed. As a result, the types of resistance have changed. Does the promotor model have to be adapted to these organizational developments?

Let us stop here. The trend is obvious: the deeper one delves into the concepts of leadership and management, the more questions about the details and development of the structures will arise.

2.2 Route 2: Extension of the promotor model

(1) The promotor model is based on the concept of resistance. Only vague theoretical concepts filed under "brakers and drivers" or "devil's advocates" or "loyal opposition" are currently available to describe the people who embody such resistance. The opposition model by Witte, Chakrabarti/Hauschildt and Gemünden/Walter objectivizes the resistance but does not personalize it. Should we not ask how - in accordance with the development of promotor structures - *opponent structures* are formed and behave? Is it not the case that the development of an innovative solution with a successful overall outcome can only be explained through the dialectic of promotors and opponents?

(2) A second corresponding model refers to the firm's partners in the innovation process. If we accept the premise that more and more innovations will be developed in future on an inter-organizational basis, the question of *corresponding "promotor structures"* at the individual cooperation partners arises. The promotor and champion constellation of a firm could also be envisaged in that of the co-operating partners, like a mirror image. The hypothesis would be that the success of the inter-organizational structure can only be secured by corresponding promotor or champion structures.

2.3 Route 3: Greater focus on supporting instruments

Research to date has been based on the tacit assumption that promotors or champions alone determine the success of innovations, without additional supporting instruments or backing. That is, of course, not the case. In actual fact, success does depend on other means, which are certainly not neglected by research. However, we know little about the reinforcing or attenuating effects with regard to the human constellation:

– Informal *information and communication* is quite certainly a major factor for success in innovation. And it is also definitely true that promotors quite clearly tend towards an informal interactive and communication style. To this extent, the two effects seem to reinforce one another. But what does informal information and communication mean in the age of electronic media? What type of informality is expected and useful? Is formal information obsolete or even disadvantageous?

– Promotors and champions are active "temporarily", for a brief period. They have the end of the innovation process in mind. What awaits them then? On the surface, this concerns the question of *incentives*, of rewards, of penalties, of all types of sanctions that firms hold in reserve for successful or unsuccessful managers. The problem is so acute because promotors and champions generally do not commit themselves because of extrinsic drivers, but intrinsically. They get involved, they commit, they are not called in or appointed. How does a firm react to such self-appointed activists? How does it deal with failed or unsuccessful innovators?

– What role does the change in *organizational or corporate culture* play in the readiness to participate and the success of promotors and champions in innovation processes? Even if we no longer accept the classical dichotomy of "mechanistic" and "organic" organization culture, Burns and Stalker (1961) nevertheless show that there is a problem here: the more mechanistic an organizational culture is, the more important power promotors are. The more organic it is, the more important process and technology promotors are. The forms of organizational change mentioned above tend more to indicate that organic forms are gaining in importance. Will the role of the power promotor become obsolete?

These three routes may initially be significant more from the theoretical point of view - but we have now learnt that there is nothing so
practical as a good theory which explains and forecasts reality. To
this extent, the contrast between theory and practice - or in our interpretation, between explaining and doing - is much less important
than is often maintained.

2.4 Route 4: Greater focus on application

(1) Innovations are projects, but not all projects are innovations.
How far can the practically orientated proposals of *project management* be used in innovation processes? The following striking point
emerges from an analysis of the literature on project management
and on innovation management:

– The project management literature emphasizes formal organizational tools for project support, such as matrix management, project controlling, network planning, cost control, information management. In contrast, it devotes much less attention to the human
  aspects of project management.

– The literature on innovation is quite different. Here, considerable
  skepticism prevails about formal organizational tools, while at the
  same time the human perspective on the management of innovation processes is emphasized.

The following question thus arises: Can the domains and overlapping areas of project management and innovation management
(promotor concept) be more sharply defined?

An analysis of the available research results prompts the following
conclusions:

– Quite clearly, the *degree of innovativeness or the complexity of
  the innovation problem* is of major importance to the human management of the process. Lechler's research findings (1997) show
  that, in particular, "strategic" projects of high complexity with a
  high degree of innovativeness can be progressed successfully by a
  troika constellation. It is notable that the formal organizational
  coordination tools of participation, planning, control, information
  and communication play a minor role. The promotor concept can
  be recognized very clearly here: the troika of promotors substitutes for formal coordination.

– The *basic notion* by which the innovation is driven is not insignificant: if the *innovation is driven by its end* (demand pull), with
new technologies being sought to meet known objectives, this is
the domain of the project management concept. It is otherwise
with *means-driven innovation* (technology push): here, a known
technology is available for which a completely new application is
being sought. In this case, the innovators must free themselves of
the earlier constraining ties and relationships. They are much
more reliant on spontaneous ideas and ad hoc creativity. In this
case, collaboration with customers or other external partners in the
innovation process play a more important role. Means-driven innovation is the domain of the promotor model.

– A third influence on the application of certain process management models is undoubtedly the *stages of the innovation process.*
Innovation processes generally have a relatively long, relatively
fuzzy lead phase in which the problem has to be defined and the
objectives set. According to all the findings available, this seems
to us to be more a domain of the promotor and champion concept.
Only after completion of the definition phase is it possible to think
about transferring the problem into a "project", i.e. of institution-
alizing it, setting a time frame on it, structuring it, giving it ac-
countability and responsibility, a formal structure for interaction.
Empirical findings advise a certain amount of caution here, however, since the leap from the "uncertain" to the more "certain"
phase is not at all clearly mapped out. Nevertheless, it would be
wrong to approach innovative projects from the start with the
toolkit of traditional project management. It would also be wrong
to continue to practice the full openness of self-management in the
establishment and realization phases, which would allow many
sections of the process to be repeated.

(2) A further practical question arises in the light of the frequent
observation that champions or promotors occur "spontaneously" and
that their emergence is not amenable to organizational intervention.
At first glance, it seems that we have to inquire resignedly whether
this is a question which may satisfy our intellectual interest in explanations, but not our practical interest in "doing".

*It is obvious that the cooperation of promotors and champions
cannot be obtained by force. But it can be facilitated.* This calls for
opportunities or nurturing conditions which improve the chances
that these creative spirits will get together. Thus, we are propagating

the *idea of "meeting points"*: a firm should create opportunities for those people who show an enthusiastic interest in a certain techno-logical or market-specific segment to meet, to become acquainted, to evaluate and appreciate one another. We are thus putting in a plea for meeting places, for informal opportunities for communication, for regular, institutionalized, open and non-hierarchical meetings.

3. Final remark

We could end our remarks here. However, the research findings in-duce us to leave the narrow world of the firm and innovation man-agement and to address further-reaching *demands to society*: if achieving innovations really depends on specific people, on their readiness to give their enthusiastic commitment, then we have to demand the particular promotion of such people by our educational system. We have seen what the key features are:

- The deliberate use of hierarchical potential: the educational sys-tem can contribute little here.

- The creative use of specialist knowledge and expertise: this is the traditional track of the academic system.

- Communicative, organizational, interactive, indeed diplomatic skills on the part of process promotors: there is a broad field open to our educational system here. Champions are unfortunately not the final product of our education. Not, or not yet?

A longitudinal examination of how champions influence others to support their projects*

Stephen K. Markham

A longitudinal examination of how champions influence others to support their projects

Stephen K. Markham

1. Overview of champion influence

If champions profoundly influence the projects they support (Schon 1963; Chakrabarti 1974; Burgelman 1983; Van de Ven, Grazman 1995; Frost, Egri 1990; Venkataraman et al. 1992) then knowing how champions contribute to projects is critical to understanding, managing, and facilitating innovation and to training others how to champion projects. Nevertheless, previous empirical research has not examined what techniques champions actually use to support their projects and what effect champions have on project perform-ance. Neither do we know the success of individual championing activities in promoting projects.

This study examines how champions influence others and what effects they have on the teams and projects they support. It includes four stages: the identification of champions and influence targets, the extent to which champions use standard influence tactics, rela-tionships between champions and targets, and the effect of champi-ons' influence attempts on project performance.

Central to the concept of the champion is the idea that champions influence others to support their projects (Schon 1963; Chakrabarti 1974; Badawy 1988; Frost, Egri 1990; Roberts 1988). Although Schon developed the concept of the champion in 1963, we still do not know the techniques champions use to support their projects. Recent studies on champions have addressed personality and influ-ence issues (Howell, Higgins 1990), political processes (Frost, Egri 1990; 1991; Markham 1998), interactions with other innovation roles (Van de Ven, Grazman 1995; Markham et al. 1991), associa-tion with new product development practices (Markham, Griffin 1998), and cultural differences in championing strategies (Shane et al. 1994). This study examines the effect of champion activities and relationships on project.

This research defines champions as people who:

1) adopt the project as their own and show personal commitment to it,

2) contribute to the project by generating support from other people in the firm, and

3) advocate the project beyond job requirement in a distinctive manner.

Champions achieve distinctiveness by accepting risk, by vigorously supporting or advocating the project, by helping the project through critical times, by overcoming opposition, or by leading coalitions.

A champion's effect may be more complicated than a simple direct effect on new product success or on project management actions. To date, few studies have demonstrated a positive champion effect on new product success. In fact, Markham and his colleagues found that champions had no direct effect on the success of projects with which they were associated (Markham et al. 1991; Markham, Griffin 1998). Yet, results of these empirical studies did find indirect champion effects through project management actions. For example, champions affect three areas: level of investment, budgets, and project termination decisions; levels of support (Markham et al. 1991); and levels of new product development (NPD) process integration and strategy innovativeness (Markham, Griffin 1998). Thus, though champions may not directly affect new product success, they may affect other people and processes in the organization.

The concept that champions influence others to support their ideas is consistent with previous research. For example, Schon (1963) said that champions "know how to use the company's informal system of relationships." Chakrabarti (1974) identified champions as "politically astute," as having "knowledge about the company," and as having "drive and aggressiveness." Roberts and Fusfeld (1981) said that champions "propose and push a new technological idea towards formal management approval." Smith and his colleagues said that a champion is the "major salesman to management" (Smith et al. 1984). These authors and others (Rothwell et al. 1974; Ettlie et al. 1984; Fischer et al. 1986; Maidique 1980) indicate that champions persuade people to support their projects rather than just performing project work themselves. These authors also suggest that the champion's form of persuasion employs a variety of influence techniques, such as selling, rationality, enthusiasm, and making personal appeals to other individuals for their assistance.

This study measures the effect of championing by the influence champions have on other people to support their projects rather than on the champions' direct work on project tasks. It tests the general hypothesis that champions affect projects through other people by developing hypotheses to examine the effects of personal relationships and champion tactics on target compliance and willingness. The study then examines the effect of target compliance and willingness on project performance.

2. Theory

2.1 Hypotheses development

Although there have been many anecdotes and observations about what champions do, there have been few attempts to systematically examine how they actually influence other people. Researchers have observed that champions push, persuade, sell, or advocate a project (Roberts, Fusfeld 1981), yet we do not find empirical evidence about how often or how successfully champions use these tactics. Given the influence of champions and the variety of tactics they are reported to use, this article examines a full range of influence activities.

Kipnis and his colleagues (Kipnis et al. 1980; 1981; 1982) derived an empirically based taxonomy of influence tactics to capture types of influence attempts used in a business setting. They developed the taxonomy because Kipnis observed rational classification schemes about how people use power, such as French and Ravens' Bases of Social Power (1959) are both contaminated and deficient (Kipnis et al. 1980). Even Raven (1974) pointed out that these rational schemes overlap each other and may not include all of the methods people actually use to influence others. Further, empirically based influence tactics are superior to rational classification schemes since rational schemes fail to account for some methods people actually use to influence others (Kipnis et al. 1980; Schilit, Locke 1982; Sullivan et al. 1990; Hinkin, Schriesheim 1988; Yukl, Falbe 1991). Kipnis and his colleagues studied hundreds of managers to construct valid and reliable measures of influence that managers employ to gain compliance. This study and Howell and Higgin's study (1990) use Kipnis' influence strategies to assess how champions influence other people to support their projects. Table 1 presents Kipnis' influence strategies. The next section develops hypotheses about tactic use and effect.

Strategy	Behavior
Reason	This strategy involves the use of facts and data to support the development of a logical argument. Sample tactic: "I explained the reason for my request."
Coalition	This strategy involves the mobilization of other people in the organization. Sample tactic: "I obtained support of coworkers to back up my request."
Ingratiation	This strategy involves the use of impression management, flattery, and the creation of goodwill. Sample tactic: "I acted very humbly while making my request."
Bargaining	This strategy involves the use of negotiation through the exchange of benefits or favors. Sample tactic: "I offered an exchange" (If you do this for me, I'll do something for you).
Assertiveness	This strategy involves the use of a direct and forceful approach. Sample tactic: "I demanded that he or she do what I request."
Higher Authority	This strategy involves gaining the support of higher levels in the organization to back up my requests. Sample tactic: "I obtained the informal support of Higher-ups."
Sanctions	This strategy involves the use of organizationally derived rewards and punishments. Sample tactic: "I threatened to give him or her an unsatisfactory performance evaluation."

Source: adapted from Kipnis and Schmidt, 1980; 1982

Table 1: Kipnis' Influence Strategies

2.2 Champion choice of influence tactics and effects on targets

The research literature indicates that some types of influence strategies are more successful at gaining compliance from influence targets. For example, people use logical argument more often and more successfully than other influence strategies (Schilit, Locke 1982; Hinkin, Schriesheim 1990; Howell, Higgins 1990). Similarly, Sullivan and his colleagues argued that people prefer to use persuasive and cooperative tactics rather than confrontive influence strategies even if they anticipate resistance (Sullivan et al. 1990). In fact, threatening tactics often result in unsuccessful influence attempts (Keys, Case 1990).

The choice to use personal power rather than positional power is positively correlated with target commitment, and persuasion and information power ought to be added to French and Ravens' list of power bases (Yukl, Falbe 1991). Pitts (1990) found that outstanding project managers, those that were successful influencers, used expert power more often than reward or legitimate power sources. Thus, we see cooperative types of influence strategies are more successful at gaining target compliance and willingness.

Researchers describe champions as politically astute and successful (Schon 1963; Chakrabarti 1974). If champions are politically astute and if cooperative tactics result in higher levels of target compliance and willingness, we should see champions use tactics that are more cooperative and less confrontational in nature. Thus, Hypothesis 1 states:

HYPOTHESIS 1:
Champions use cooperative influence tactics more often than confrontive tactics.

If champions use cooperative influence tactics and cooperative tactics are related to higher levels of compliance and willingness, then we should see higher levels of compliance from targets. Thus, Hypothesis 2 states:

HYPOTHESIS 2:
The more champions use cooperative tactics, the more successfully they influence their targets.

Hypotheses 1 and 2 suggest that champions use cooperative tactics because they lead to successful influence. The research literature suggests pre-existing relationships between influencer and targets likely affect which tactics are used (Sullivan et al. 1990).

2.3 Champion-target relationship and the choice of tactics

Research literature identifies a number of variables associated with building relationships (Yukl, Falbe 1991). The choice of influence tactics according to rational models of power (French, Raven 1959) suggest that once champions know the source of power that works best with their targets, they know the right tactics to employ. For example, a reward power base suggests that champions use rewards to influence others. Similarly, a coercive power base results in threats and punitive actions.

The power and influence literature, however, does not support this theory (Kipnis, Schmidt 1982). Dillard and Burgoon (1985) found that power has little effect on a champion's choice of tactics. Later, Burgoon et al. (1987) found dominance has little or no impact on compliance. In fact, Hinkin and Schriesheim (1988) found that assertiveness actually lowers subordinates' compliance. Miller (1982) found the use of power by itself has no impact on choosing an influence strategy, but power did interact with the quality of the champion and target's relationship. Finally, supporting the research that power is not related to tactic choice, Vecciho and Sussmann (1991) found that tactic use across different levels in the organization did not differ.

Although researchers have theorized organizational power might affect tactic choice, if champions are politically astute, they will not use tactics, such as organizational power, to influence others. Thus, a champion's organizational power will not serve as a predictor of his or her tactic choices. A major reason that influence strategy choice is important to champions is that certain influence attempts, such as the use of organizational power, may strain relationships needed for future assistance. Power and influence are different phenomena, and factors other than power impact a champion's choice of influence strategies.

Further, the choice of tactics may not entirely depend on a particular champion style or general preference. George Graen and his colleagues (Graen 1975; Graen, Cashman 1975; Graen, Scandura

1987) found that the quality of personal relationships between leaders and followers is a major determinant in choosing influence techniques. Seers (1989) later found the quality of personal relationships determines influence activities among team members. Therefore, champions in high quality relationships choose tactics that preserve the relationship between the champion and the target. This theory suggests that champions use cooperative rather than confrontive tactics to influence targets with whom they have high quality relationships. Hypothesis 3 states:

HYPOTHESIS 3:
The higher the quality of relationship that champions have with their targets of influence, the more they use cooperative rather than confrontational tactics.

Therefore, this study hypothesizes personal relationships between champions and targets, rather than organizational power, affects tactic choice.

2.4 Quality of relationship and level of influence

Not only will the quality of relationship affect the choice of influence tactics, it will also directly affect the targets' compliance and willingness. The quality of relationship affects the compliance and willingness of in-group and out-group members (Graen 1976). Additionally, the level of compliance and willingness of team members relies on the quality of relationships within the team (Seers 1989).

High quality relationships are necessary to influence other people (Liden, Mitchell 1988). Keys and Case (1990) found that the managers' formal authority is often small and influential managers maintain good relationships. Positive relationships depend on quality communications (Drake, Moberg 1986). The more an influencer works with a target, the easier the exchange process between them (Schilit 1987). Conversely, adversarial relationships tended to reduce one's ability to influence (Chacko 1990). Thus, we predict that champions will be more influential with targets they have good relationships with. Hypothesis 4 states:

HYPOTHESIS 4:
The better the relationship between champions and their targets, the more successful they are at influencing those targets.

Therefore, this study hypothesizes that the champion's proper choice of tactics and quality of relationship result in higher levels of target compliance and willingness.

2.5 Effect of target compliance and willingness on project performance

Although researchers generally accept that champions have strong positive influence on the projects they support, empirical evidence supporting this notion is scarce (Fischer et al. 1986). In examining what champions do, researchers have not yet established how champions influence projects. Although Howell and Higgins (1990) examined champions' influence attempts, they did not examine the champions' impact on the project. Markham and his colleagues found that champions influence management decisions, including resources, and termination, and that these management decisions affected final project performance. They, however, did not study target influence attempts (Markham et al. 1991). Markham and Griffin (1998) found relationships between the presence of champions and best practices in new product development and that those practices affect project performance, but again they did not examine champion influence behavior. This study examines the relationship between what champions actually do and their impact on the projects they support.

The methods champions use to influence projects may not be as straight forward or direct as researchers previously thought. Therefore, this study does not examine a champion's own work on a project. Rather, consistent with the definition of a champion, it examines the champion's impact on a project through influencing others to support it. The hypotheses of champion influence proposed in this paper measure the willingness and compliance of the champions' targets of influence, rather than the effects of the champions' own work on a project.

Researchers also suggest an indirect relationship between influencer activity and performance. Eisenhardt and Bourgeois (1988) found autocratic use of power leads to negative political behavior, which, in turn, managers cited as the cause of poor performance. Similarly, Hinkin and Schriescheim (1988) found that incorrect influence tactics lowers compliance and thus diminishes subsequent performance. Kipnis et al. (1979) observed that the use of power and target compliance is a floating issue, and that compliance moves from one issue to another. He noted that if a manager impairs employees in one area of their work, they may attempt to impair the manager in another area of work. Champions seem to understand and avoid this metamorphic effect of power. They nurture their targets of influence so that the targets will not refuse their requests for the extra effort or favors that they need for their projects.

Therefore, by gaining support from other people, champions help contribute what is needed for the project to progress. If champions successfully influence their targets in such a manner that the targets' positive response helps the project, one expects to see higher levels of project performance. Therefore, Hypothesis 5 states:

HYPOTHESIS 5:
The more successfully champions influence their targets, the better the champions' projects perform.

Through the use of champion and power literature, these five hypotheses examine the tactics champions use and the impact of the champion and target's relationship on the choice of the tactics (Figure 1). The hypotheses also examine the impact of the champion and target's relationship on target behavior and how influencing affects project performance. The next section examines the methods used to test these hypotheses. The resulting insight into how champions influence others and how that influence translates into project performance has implications for researchers and managers.

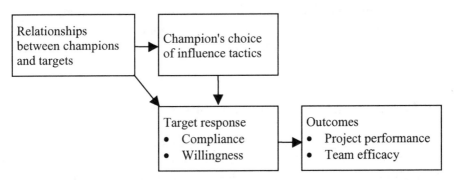

Note: the effects of champions is through their targets

Figure 1: Effects of champion relationship and choice of influence tactics on project outcomes

3. Methods

3.1 Sample

The respondents consisted of fifty-three champions of innovation projects and team members. Management from four Fortune 500 companies identified seventy-eight people for this study. Twenty-five individuals did not meet the definitional criteria of a champion described below and did not remain in the study. Fifty-three champions completed the time one questionnaire and forty-six champions completed the time two questionnaire about influence techniques. These were all people who evolved into the role of the champion, no one that was assigned to be a project champion by management was allowed into the study. On average champions were 38.2 years old and had been with their present organization 6.5 yeas and in their present position 3.6 years. They averaged 17.2 years of work experience and 55% had graduate degrees. Eighty percent of the champions were male. Two of the females expressed reservations about being identified as champions, they preferred to think of themselves as team members. None of the males expressed this reservation. Champions typically had manager or director titles.

The champions identified 189 team members, an average of 3.2 per project. 138 team members (73%) responded to the time one questionnaire. The same number of team members (138) returned time two questionnaires.

3.2 Definition of the champion construct

Schon (1963) originally described a champion as a person "willing to put himself [herself] on the line for an idea of doubtful success." According to Schon, champions defend a project with a heroic quality even if it means "professional suicide." Other champion descriptions range from simply project supporters (Chakrabarti 1974) to people in informal roles (Badawy 1988) who act as risk-taking charismatic leaders (Howell, Higgins 1990).

Most descriptions appear to identify three types of champion activities (Schon 1963; Chakrabarti 1974; Rothwell et al. 1974; Maidique 1980; Roberts, Fusfeld 1981; Fischer et al. 1986; Chakrabarti, Hauschildt 1989; Day 1994). A review of the champion literature and pilot interviews with the 30 teams provide a rich description of championing activities. First, the champion in a personal way adopts an innovation during development. Second, the champion contributes to the development of the innovation by promoting it within the organization by materially and emotionally supporting the development process. Third, a number of these authors imply that the champions' behavior is distinctive in sponsoring the innovation because they incur risk associated with visibly advocating a project, overcoming opposition, and trying to enlist greater levels of organizational support.

3.3 Procedures

This study incorporates a cascading two-stage longitudinal design. The first stage, at time one, requires the identification of champions and team members for study and the collection of influence techniques champions use on their targets. The second stage measured influence outcomes and tactics used at time two.

When a firm agreed to participate in this study, its management identified a vice president or general manager to work with the research project. These liaisons helped identify champions that were not formally assigned as champions, innovation projects, and personnel, and provided company and project information. Four criteria determined a project's suitability for this study. First, after careful instructions about this study's definition of a champion, the liaison identified a person that he or she felt was generally accepted as the champion for the project. Second, the liaison chose projects with identifiable teams to allow collection of performance data apart from the champion. Third, the liaison chose ongoing projects expected to continue for at least two additional months to follow champions in real time. Fourth, the liaison chose a champion presently involved enough with the project to report influence activities over a two-to four-month period. Innovation projects include both product development projects and technology adoption projects. The championing process appears stable between these types of projects (Lawless, Price 1992). Similarly, Damanpour (1991) found that the dimensions of the innovation process were stable across different types of innovation projects.

For this research, a liaison first identified championed projects. Then, the liaison identified the project champions, the champions identified the team members and targets, and the team members, in turn, validated the nomination of the champion. Then, the researcher interviewed the champions for the study. The results of a structured interview with the person nominated as a champion determined if he or she met the definitional criteria developed above (personally adopt the project, generate support from others, distinctively advocate the project). To be considered champions for this study, the nominees met all three of the following conditions:

1) nomination by senior management as a person acting as a champion,

2) indication in interview that they indeed function as champions as specified in the definition, and

3) identification by at least two of the team members that they function as champions for specific projects.

In addition, the champions could not be assigned to be a champion. In interviews, team members confirmed if a champion was present for a given project and, if so the identification of the champion. The interviewer did not reveal to the team members the champion's identity or inform the team members that a champion had named them as project team members. Thus, both senior management and team members independently identified the champions for this study. The researcher screened the champions and eliminated twenty-five nominees from the original sample. The eliminated individuals included nominees not identified as project champions by at least two team members or who did not indicate in their interviews that they participated as a champion.

The last actors that identified were the targets of influence. The researcher identified targets individually by asking the champions to identify one-to-three critical targets. All fifty-three champions identified three targets each. The researcher limited a champion's number of targets to three to maintain the amount of data requested from each champion at a reasonable level. If the champion identified the target as a group of people, such as a marketing department, the researcher asked the champion to identify the person who served as the primary decision-maker for the project. At time two, the researcher again asked champions how they influenced the three targets they named in time one. This procedure produced a comparable measure of the tactics used with the same target over time.

3.4 Measures

The researcher collected data from different respondents and in both time periods. All variables contained multiple measures either across time or from different respondents, and the researcher analyzed the measures by principle factor analysis to assure comparable factor structures across the measures. The study assured dimensionally by retaining factors that met the minimum eigen value (Kaiser 1974) and scree test criteria (Tucker et al. 1969) across the different measures. Also, the study determined item appropriateness or item retention in each variable by minimum factor loading criterion (Kim, Mueller 1978). As a result, the researcher retained items only if they met measurement criteria across all available sources. Most measures met minimum alpha criteria ($>.70$) for both time periods, with many measures above .80, and some above .90. The target compliance variables were in the mid .60s.

One reason for collecting data from multiple measures is that an influence strategy may encompass a lagged effect. Collecting data on the use of influence and target reaction at one point in time may not allow the full effect of the assessed influence. Also, collecting data at different times and from different respondents reduces same source bias.

Table 2 indicates the respondents for each variable and when they responded.

Variables	Respondent	Time Responded
Influence Strategies	Champion	Time 1, Time 2
Target Compliance	Champion	Time 1, Time 2
Target Willingness	Champion	Time 1, Time 2
Champion-Target Relationship	Champion	Time 1
Technical Performance	Team Members	Time 1, Time 2
Expected Financial Performance	Team Members	Time 1, Time 2
Goal Attainment	Team Members	Time 1, Time 2
Project Efficiency	Team Members	Time 1, Time 2
Team Efficacy	Team Members	Time 2

Table 2: Variables, respondents, and response times

3.4.1 Influence strategies

The champions responded to open-ended questions about what they wanted to accomplish and how they accomplished it. The researcher then asked the champions to rate on a five-point scale (1 = none; 5 = extensive) how extensively they used eight different strategies to influence the people they identified as targets at both time one and time two. The influence strategies used in this study are adaptations of Kipnis and Schmidt's strategies (1982). The eight strategies include rational appeals, coalition formation, ingratiating, bargaining, assertion, appeals to higher authority, coercion, and clandestine activities (See table 1.). The assignment of strategies as either co-operative or confrontational closely followed factors derived by Falbo and Peplau's dimensions (1980). Cooperative influence strategies, in which the champion includes the target in defining the solution, are rational, ingratiating, and bargaining. The confrontational strategies, in which the champion forces a solution on the target, are coalition, assertiveness, higher authority, and sanctions.

3.4.2 Relationship

The researcher assessed the relationships between the champions and each target of influence by using an adaptation of Seers' Team Member Exchange (TMX) concept (1989). Seers adapted Graen's Leader Member Exchange (LMX) concept to assess the relationship between team members. This research used TMX instead of LMX because pilot interviews with product development teams revealed that issues between champions and targets were more closely related to team member relationships than leader-subordinate relationships. This study uses nine of Seers' items to assess the relationship between champion and targets.

3.4.3 Target compliance and willingness

Champions served as the sole respondents to questions about target behavior. In this study, the researcher never approached targets for data collection because the projects were ongoing and such an effort might broach confidentiality, alter project dynamics, or contaminate the data. Also, the champions were uneasy about targets being questioned about the success of the champions' influence activities. The study, then, assessed two aspects of target behavior: the level of compliance and the willingness to comply with champions' requests. Because the study measured compliance at the target level, the researcher analyzed each target for each time; thus, there are 159 observations in the data, with tactics nested in champion. The researcher assessed the level of target compliance using three items. A sample target compliance item was "Did this person comply with your requests for help on this project? (1 = rarely, 5 = often)." The researcher then assessed the level of target willingness with six items. A sample target willingness item was "The target has become more committed to seeing the project through. (1 = disagree, 5 = agree)."

3.4.4 Project performance

This study included performance measures at the project level rather than at the individual interaction level; therefore, it uses the average of the team member responses. The researcher adapted sixteen items assessing project performance from Green's et al. (1985) multidimensional performance measure At time one and time two, the researcher administered the project performance scale to the team members only. Three items assessed whether the project continued to perform above or below technical expectations. Four items assessed the expected financial impact of the project on the firm. Four items assessed how efficiently the project used resources. Three items assessed whether the project was above or below expectations of achieving project goals. The researcher eliminated two items because they did not meet minimum factor loading criterion.

3.4.5 Team efficacy

An important part of project performance is whether the team increases its capacity to work on similar problems in the future. The researcher devised nine team-efficacy (TE) items according to Bandura's concept of self-efficacy (Bandura 1986; 1988) and retained all items for analysis. A sample TE item was "How confident are you that the team will develop new knowledge useful to the firm? (1 = low – 5 = high)."

4. Results

In the interviews, the researcher asked each champion candidate why he or she was involved in the project. Most of the champions answered this question by first referring to the "the project" as "my project". In nearly every case, answers included the word "passion". The champions obviously took personal ownership of their projects and exhibited a high level of attachment that often translated into protection. The researcher asked each champion to relate the project's history and how it progressed to its current development phase. Stories universally included episodes of extraordinary effort and risk on the champion's part. Although there is a natural bias to focus on one's own contribution, these projects were all in process, and the success of many projects was not guaranteed. Thus, while the champions admitted they had accomplished considerable work on their projects, they displayed little inclination to take full credit for them.

The most common story told by champions and team members about important project contributions focused on champions' trying to influence other people to provide critical assistance to the project. Conversations with team members revealed that team members not only recognized the champion, but also that the champion was taking a risk by advocating the project. Comments from team members included, "If the project doesn't work out, Jack will probably get sacked," or "Carol has a lot riding on this project." Champions universally recognized their risk, but they also described passion for what they were doing and motivation to complete the project. Yet, champions are affected by the risks they take. They often related stories about the personal toll projects were taking on their health, and their professional and personal lives.

Hypothesis 1 states that champions prefer to use cooperative rather than confrontational tactics. The mean use of cooperative influence strategies was 3.05 for time one and 3.00 for time two. The mean averages for confrontive tactics were lower: 2.75 at time one and 2.68 at time two. Testing the hypothesis included two methods: averaging the cooperative tactics (reason, ingratiating, bargaining) and confrontive tactics (coalition, assertiveness, higher authority) and running a t-test between means. The researcher omitted the sanction tactic from all analyses because so few champions reported using it. Results indicate support for the hypothesis that champions use cooperative tactics more often than confrontive tactics at time one (t = 10.85, n = 238, p <.0001) and time two (t = 13.10, n = 238, p <.0001).

Table 3 reveals that, over time, champions tend to use tactics at the same level. An exception occurs when champions make appeals to higher authority figures, this activity significantly decreased over time.

Variable	N	Time 1 Mean	Time 2 Mean	T Value	Pearson r
Champion Tactics					
1. Rational	138	3.96	3.93	-.94	.42***
2. Coalition	138	3.40	3.44	.68	.23**
3. Ingratiation	138	3.17	3.13	-.06	.34***
4. Bargaining	138	2.01	1.94	-.55	.38***
5. Assertion	138	1.65	1.68	.41	.31***
6. Higher Authority	138	3.19	2.93	-2.68**	.49***
Target Behaviors					
7. Compliance	138	3.91	3.88	-.62	.70***
8. Willingness	138	3.84	3.55	-4.33***	.52***
Project Performance					
9. Support	238	3.54	3.66	1.93*	.36***
10. Resources	238	3.12	3.28	2.20*	.13*
11. Wellness	238	3.19	3.37	2.46**	.37**
12. Technical	238	3.54	3.79	3.94***	.16**
13. Financial	238	3.64	3.68	.42	.21**
14. Goals Attain	238	2.84	3.21	5.18***	.23***
15. Efficiency	238	3.35	3.45	1.64+	.23***
16. Team Efficacy	238	5.67	6.20	2.98***	.06

*** = p <.001, ** = p <.01, * = p <.05, + = p <.10

Table 3: Means, T-tests, and correlations between repeated measures

Hypothesis 2 states that cooperative strategies generate more compliance and willingness than confrontive strategies. The researcher tested this hypothesis by using a multiple OLS regression model. Controlling for quality of relationship, the equation regressed compliance and willingness at time two on influence strategies at time one. Since the champion was the sole respondent for both tactics and target behaviors, using measures from different times reduced same source bias. Secondly, to further reduce this same source bias, the researcher asked questions about influence tactics at time one and questions about target behaviors at time two. This questioning method made it difficult for champions to remember their previous responses and increased the likelihood that the research could determine the effects of influence tactics at time one and the effects of target compliance and willingness at time two. Yet, results did not support the second hypothesis (See table 4.). Neither cooperative nor confrontational tactics increased the likelihood of more target compliance or willingness. The only significant tactic effect observed was that rational tactics have a negative effect on compliance (See table 4.). This finding counters previous findings that rational tactics are the most efficacious (Howell, Higgins 1990; Kipnis, Schmidt 1988).

Variables Time 1	Compliance Time 2	Willingness Time 2
Rational	-.12+[a]	-.07
Coalition	.07	-.05
Ingratiation	.05	.04
Bargaining	.05	.05
Assertion	.11	.07
Higher Authority	.03	.08
Quality of Relationship	.47***	.49***
R2	.29	.29
F	7.63	7.67
P	.0001	.0001

a = beta estimates; + = $p = .06$; *** = $p < .0001$

Table 4: Effects of tactic and relationship on target compliance and willingness

To explore further the relationship between tactic use and compliance, post hoc analyses regressed target behaviors on only the cooperative and confrontational sets of tactics for both time one and time two tactics. The results of these analyses did not support the hypothesis either.

Hypothesis 3 states that higher quality relationships result in the champions' using more cooperative rather than confrontational influence strategies. The researcher tested this hypothesis by using OLS regression. The longitudinal data allows testing the effect of relationships over time. The equation regressed quality of relationship on champions' choice of tactics used at time two. The data did not support the hypothesis ($F(1,137)=1.24$, $p=.29$). The quality of relationship between the champion and the target did not modify the champion's choice of tactics. To further assess the effect of relationship on tactic choice, the researcher conducted post hoc analyses regressing quality of relationship on time one tactics. This additional analysis also failed to support the hypothesis.

Hypothesis 4 states that the higher the quality of relationship between the champion and the influence target, the more successfully the champion influences the target. To test this hypothesis, the researcher used the OLS regression described for Hypothesis 2. The researcher omitted the sanction tactic from all analyses because so few champions reported using it. Table 4 reveals that the betas for quality of relationship have a powerful impact on target compliance and willingness.

Hypothesis 5 states that the more successfully champions influence their targets, the better the champions' projects perform. This study measured performance on three levels:

1) building the team's confidence,

2) developing project support and resources, and

3) meeting technical, efficiency, and financial objectives.

Table 3 reveals that, over time, the project increased in all performance aspects (except financial) and that performance levels were high. Multiple OLS regressions, however, failed to find a relationship between target compliance and performance. The equations regressed time two performance variables on time two compliance and willingness. This hypothesis was partly supported because target compliance and willingness predicted team efficacy ($F(1,137)=3.29$, $p=.04$). Nevertheless, even if the targets willingly complied with the champions' requests, the compliance did not translate into project support, resources, or in meeting technical, financial and efficiency objectives. Yet, even though the team did not discern resources or organizational support, the champions did. Follow-up analyses reveal very different results through examining champion responses to project support and resources. According to the champions, target compliance resulted in the project enjoying both support (F $(1,137)$ =5.02, p =.007) and resources (F $(1,137)$ = 8.45, p =.0003). Other analyses also failed to establish a relationship between tactics or quality of relationship and performance.

To summarize the results, based on these four firms, champions use cooperative rather than confrontive influence tactics. Nevertheless, the champions' choice of tactics does not have an effect on the level of target compliance and willingness to participate in the project. Neither does the quality of relationship between champions and their targets modify the champions' choice of influence tactics. Overall, the tactics that champions use do not appear to play an important role in these projects. Yet, it appears that champions substantially influence their targets' behavior if they have good personal relationships with the targets. Finally, although champions do have an impact on their targets, the team members do not recognize heightened project performance. Champions, however, see their targets' actions resulting in higher levels of project support and resources.

5. Discussion

The research only partially supports the general hypothesis that champions affect projects through influencing other people. This study explores two commonly accepted influence methods, personal relationships and influence tactics, that people, including champions, might use to influence other people to support their projects. The study also presents the effect champions' influence has on the team and project performance according to the team. Champions do positively influence target activities though their relationships but not through commonly used influence tactics. Additionally, successfully influencing others did not affect project performance.

These results support previous empirical research findings that champions have a limited effect on the projects they support. Markham and his colleagues found that champions have important effects on management practices, including project support, budget, continuation, NPD process integration, and innovative strategies, but not on final performance (Markham et al. 1991; Markham, Griffin 1998). Similarly, this study found that champions impact individual behavior through relationships, but it did not detect a direct impact on projects. These results converge to indicate that researchers and managers fundamentally do not understand the championing process.

Together with other studies, these findings call into question the popular notion that champions have a positive impact on project performance. This study also raises doubts that targets receive well the champions' influencing activities. Although many authors have ascribed positive outcomes to champions (Schon 1963; Chakrabarti 1974; Markham et al. 1991; Day 1994), their retrospective data do not correspond to the realities of ongoing projects.

These data may or may not suggest that champions are ineffective, but they do indicate that the traditional image of champions' having a positive effect on project performance is still speculative. This study examined ongoing projects; however, champions may conduct much of their work before the organization actually funds or sanctions the project. Thus, the champion's major contribution may occur at the project's inception or in the Fuzzy Front End. The influence of champions during project development may be less than it was during inception, and it even may be described as detrimental. The finding suggests that champions may not be as influential at all stages of a project.

The methodology of examining the response of only a few targets at two times may not accurately measure the amount of influence champions actually exert on behalf of their projects. All fifty-three champions named the maximum number (3) of targets. This study also included weekly telephone interviews with champions where they indicated a far larger range of targets than the questionnaires could capture. In these periodic interviews, they also indicated a high frequency of influence attempts, often more than ten a day. The methodology used here is not sensitive to that level of activity and, therefore, may not detect a performance change due to the relatively few influence attempts captured in these data.

Most of the champions in this study felt they were locked into a "life and death struggle" for the project, but that they were winning the struggle. It could be that initiating and keeping projects alive is the major contribution that team members may not observe. During their development phases, all projects may appear as though they are performing poorly and that resources for the project are scarce. Additionally, a team may not notice or report a champion's successfully saving a project from termination. Interestingly, the champions' views of the projects are much more positive than team members. Champions appear to see the positive even when others do not. Team members may not regard this tenacious optimism as wise until or unless the project succeeds.

Schon (1963) suggested that projects either find champions or die. It may be too much to ask of champions to keep projects alive and to guarantee positive performances. Champions must accept a great deal of risk and risk their careers for a project they cannot control. These data even suggest that the more champions try to influence others to support the project, the more other people may become less supportive (See table 3.). In this case, champions could even negatively influence project performance and the team. Thus, it is likely that team members can see and value champions' contributions only after the project becomes secure.

This study examines the activities of champions in ongoing projects. The data suggests a fairly negative experience for all parties. Nevertheless, performance did improve significantly over time. There may be several explanations: First, champions may use other methods to transfer their influence to project performance. For example, project performances at time one could be strong predictors of which tactics champions use. Also, champion behavior may directly predict project outcomes. Perhaps other variables, like organizational constraints that prohibit target compliance, need inclusion. Another idea is that if the champions' and/or the targets' levels in the organization predict both tactic usage and target behavior, then the interaction of tactics and the quality of exchange could predict target behavior. Also, a project's age, size, cost, potential, or company experience could function as important predictors. Perhaps segmenting the data by level of target compliance or project performance might explain these unexpected findings. Yet, in post hoc analyses, the researcher tested each of these variables and analyses and many other variables and model configurations, but the post hoc analyses did not yield results any different than the original hypotheses. The results presented here were extremely robust in the face of alternative models and use of data across time. This study supported Damanpour's (1991) meta-analysis finding that the innovation process is stable across various dimensions.

There are a number of important innovation variables not measured in this study. For example, the presence of other role holders such as a gatekeeper as not measured. Similarly the use of stage gate systems or multidisciplinary teams were not considered. These and other variable may influence these results through interaction effects not tested in these analyses.

Another possible reason this study produced unexpected findings may be due to inadequate measures. The measures may not have been able to capture the champions' real use of tactics, the targets' behavior, or the projects' performances. When interpreting these counterintuitive results, the researcher must carefully consider developing measures. Discussions with some of the champions, which the researcher conducted as a check on these results, did not present the champions with a counterintuitive picture of the experiences they underwent to support their projects. All of the champions appreciated the fact that somebody finally recognized that supporting a project was difficult, demanding, and dangerous. The researcher retained variable items only if they loaded correctly at both time periods. In fact, this study's variables survived rigorous test and retest criteria. All team variables used both multiple respondents and repeated measures.

A drawback of the methodology is the fact that champions were the sole respondents about their targets' compliance and willingness. Given the ongoing nature of the study, it was untenable to ask the targets how much a specific person tried to influence them and how well they had responded to influence attempts. Since the champions were influencing the targets, one might expect that they would tend to inflate their targets' responses, and perhaps they did. Nevertheless, target compliance and, especially, willingness decreased over time even as project performance increased. This trend tends to increase confidence in the champions' responses and in the interpretation that projects are tense and tumultuous.

Another possible reason for these results is that this study did not find authentic champions. This study used multiple criteria from superiors, subordinates, and interviewers to identify champions. No other reasonable screens seemed suitable for the selection process, especially when interviews with other company employees revealed that they regarded the identified individuals as champions.

6. Implications

6.1 Research implications

This study informs the influence literature that longitudinal results revealed that people tend to use the same tactics over time. These data also tie tactic use to actual target behavior. Although one might expect tactics like assertion to be perceived negatively by a target, assertion, in fact, was not negatively interpreted when the researcher assessed actual target behavior. Similarly, other tactic studies found that people used rational tactics most frequently (Kipnis, Schmidt 1982; Howell, Higgins 1990). This study replicated those results, but it found rational tactics to be negative in their effects on performance. Research examining the longitudinal use and impact of tactics clearly is necessary. Establishing the impact different types of tactics have on influence targets is another area requiring additional research.

These data suggest that champions may not always deliver cooperative supporters, confident teams, or high performing projects. These data and other recent empirical research suggest we have more to understand about the championing phenomenon. Future research should address the longitudinal nature of the championing process and the mechanisms by which champions contribute to projects.

This study produced unexpected results that call into question the nature and contributions of champion behavior. Future research needs to establish exactly what role champions play. Topics for this research might include assessing champion existence and activities early in projects, career implications of championing, reward systems for champions and teams, assessing champions' contributions to projects, refining a definition and the domain of champions, determining other roles relevant to champions and how they interact, determining a champion's motivation to act, and assessing the cultural and gender dimension of championing. To fully understand the effect champions have on projects, we need research examining the full range of championing activities on the projects. Finally, this research indicates that we should focus on the goals champions are trying to accomplish to give us a broader understanding than we have gained from assessments of influence episodes.

6.2 Managerial implications

The championing process is not a neat, clean exercise in management. Promoting a project involves influencing many people who may not wish to be influenced. This situation may lead to tension and conflicts even if the champion and targets of influence have positive preexisting relationships. The contribution and expertise of champions may not be apparent during the project. In fact, champions may not possess the expertise necessary to make good decisions all the time and managers might establish guidelines and checks on champions. For example, introducing an antagonist may help the champion and team to refine their thinking about their projects (Markham et al. 1991). Not only do champions need supervision and monitoring, but they may also need help. Managers can assist champions who have difficulty directing others to contribute positively to projects. For example, since relationships are important to target influence, training champions in interpersonal relations may prove useful. Similarly, basic management techniques of time, meeting, and stress management could provide champions with tools for successful target influence.

Managers cannot take a hands-off approach to a project just because it has a champion. If champions take a renegade approach (Kipnis, Schmidt 1988), they could alter the company's research and development portfolio. Such diversion of resources can be wasteful and ignores the strategic direction of the firm. In either case, management must actively direct the selection and priority of projects. The danger in recognizing that champions need supervision and assistance is that management may become so involved in specifying the innovation process that creativity is stifled.

Traditionally, managers believe that a champion's enthusiasm and drive is sufficient to make projects succeed. Nevertheless, the use of influence tactics in this study, especially rational ones, seems to have a detrimental effect on targets and projects. Yet, these results do suggest a much more effective scenario. Since a high quality of exchange between a champion and his or her targets of influence results in compliance and willingness, management should assign someone with an extensive network to guide the champion. Ideally, this person should have relationships with people who are vital to the project. Enthusiastic technical people willing to use influence tactics without an established network may easily fail in their attempts to work with the champion and the project team.

Firms may provide training for many employees, including the champions, in how to establish high quality relationships with co-workers. Without exception, all champions felt strongly that they were making a personal commitment to the project. For this reason, formalization of the champion role is unlikely to produce favorable results.

Even though these results suggest revisiting champion legends, the role of the champion is still vital and interesting across different types of innovation projects. Findings that call into question common assumptions about a phenomenon enrich rather than detract from our understanding.

Acknowledgements

This research was conducted with the support of the Product Development and Management Association and the Center for Manufacturing Management Enterprises. Reviewers are also thanked for many helpful comments and suggestions.

Opposition to innovations – destructive or constructive ?

Jürgen Hauschildt

Opposition to innovations - destructive or constructive?

Jürgen Hauschildt[1]

1. Opposition as a typical attitude to innovations

1.1 1912: Even Schumpeter refers to "persistent resistance to change"

Innovations are desired. Innovations are given support. No politician, entrepreneur, top manager, professional association official, no economic theorist ever comes out openly against them. Those involved in and affected by new ideas in these institutions will also constantly affirm and re-affirm their readiness to innovate. And yet - these affirmations are often nothing but lip service. To many people, innovations are synonymous with disruption, nuisance, revolution and, in many cases, with nothing more than pointless turbulence. *Innovators have to reckon with opponents*, and not only in hidebound companies. Even "the most up-to-date firm has a persistent resistance to change", as Schumpeter found as early as 1912 (Schumpeter 1912, p. 108).

Resistance (the terms "resistance" and "opposition" are used synonymously here) to innovation is thus by no means unknown. Witte, referring to Sandig's "drivers" and "brakers" (1933, p. 349 ff.), mentions opponents along with promotors of innovation as the force and counter-force in the decision-making process (Witte, 1976, p. 319 ff.). Schmeisser (1984, p. 67 ff.), Zaltman et al. (1984, p. 85 ff.),

---

[1] German version:
Widerstand gegen Innovationen – destruktiv oder konstruktiv? To appear in: Zeitschrift für Betriebswirtschaft 1999.

Staudt (1985, p. 355 ff.) and Bitzer/Poppe (1993, p. 318) have mapped out highly complex pictures of opposition. The biographies of innovators provide impressive evidence of opposition to innovation (Hauschildt 1997, p. 125 ff.). Klöter (1997, p. 150 ff.) classifies resistance according to four criteria, making a distinction between "destructive" and "constructive" opposition.

Consulting other academic disciplines on the subject is also a valuable exercise: the interdisciplinary study conducted by Böhnisch (1979, p. 25 ff.) provides a wealth of psychological reasons for resistance by the individual to significant change. Group modes of behavior appear to reinforce this opposition in many instances. The most comprehensive socio-psychological model in this regard was presented by Janis (1982, p. 244) under the title of "groupthink". For a concept based more in the political sciences, see Frost/Egri (1991, p. 229 ff.).

By integrating the wide variety of models to be found in the literature, a multi-level resistance model can be developed. Opposition is always associated with at least one person, whose attitude is revealed in different ways:

- On the surface, "rational" arguments against the innovation are presented. These are mainly of a technical, economic, legal and ecological nature.
- Beneath this surface lie the deeper reasons for opposition, which can be characterized as "barriers of ignorance" and "barriers of unwillingness" (Witte 1973, p. 6 ff.).

These distinctions are particularly important to empirical research, because it seems likely that these different levels cannot be investigated by the same method.

1.2 Ambivalent assessment of opposition

Opposition to innovations is generally regarded as negative. The main reason for this assessment is that resistance to innovations results from specific conflicts: innovations regularly lead to conflicts of perception and knowledge. They bring conflicts of motives to light and are generally associated with conflicts of distribution. If different corporate units are involved, conflicts of role and departmental responsibility arise. Ultimately, different parties' claims to power are behind many of these conflicts: thus, innovations spark off power struggles.

And since innovations always tread new ground, no familiar hierarchical and non-hierarchical conflict handling mechanisms are available. People who avoid conflicts and see them primarily as crises of co-ordination that take away from efficiency will associate opposition with nothing but negatives (Steinle 1993, col. 2201).

But a different point of view credits opposition with positive effects: innovators without opponents lapse into an uncritical state of hectic activity, in which too many projects are started but not completed (Nolan 1989, p. 16 ff.). Simple rejection of conflicts overlooks an important fact: the effort to resolve conflicts inspires the creative imagination; "discrepancies, tensions and conflicts become a prerequisite for innovation and change" (Steinle 1993, col. 2201). In the process of canonization, the "devil's advocate" is deliberately institutionalized in order to reveal the weaknesses in the arguments of "God's advocate" through opposition (Schwenk 1990, p. 161 ff.; Valacich/Schwenk 1995, p. 369 ff.). And finally: opposition is a deep-rooted democratic tool for controlling unilateral action by the ruling party. Moreover, empirical observation shows that a "loyal opposition" is to be found much more often that would be supposed (Markham et al. 1991, p. 235).

In short: *conflicts appear to be not only the effect, but also the cause of innovations.*

1.3 Questions addressed in this research

It was this contradictory assessment of opposition to innovation that led to the present study. First of all, we have to say that empirically based knowledge about the type, intensity and impact of opposition in innovation processes has only just started to emerge. Only the empirical studies by Bitzer (1990, p. 58 ff.), Markham et al. (1991, p. 217 ff.) and Gierschner (1991, p. 289 ff.) provide initial, but heterogeneous indications of the existence and origin of opponents. To this extent, this study's mission is primarily heuristic. Specifically, the following questions about the existence of opposition are addressed:

- To what *extent* does opposition to innovations actually occur?
- What *arguments* does the opposition use?
- Can the *intensity* of the opposition be derived from the arguments used by opponents?
- Is opposition *overt or covert*?
- Is opposition always *assessed* as negative?

Causes and effects can only be investigated once these questions are answered, when the existence and variance of opposition to innovations are determined. Our survey was based on the following model (Fig. 1).

The core of the analysis is the relationship between resistance and the success of the innovation. In order to highlight the effects, we will first take the negative view of opposition, in which resistance is thought to aim at preventing, delaying or changing the innovation. These influences reduce effectiveness (here: degree of innovativeness, technical and economic success) and efficiency (here: adherence to time and cost objectives). Thus, the core hypothesis is:

HYPOTHESIS 1:
As opposition to innovation increases, its effectiveness and efficiency will decline.

This fundamental relationship will be modified with regard to the following situational influences:

Opposition is fostered by complexity. As complexity increases - i.e. as the number of individuals passively affected by and actively engaged in the innovation increases, as uncertainty about the situation and the alternatives grows, as the problem's outlines become increasingly blurred, as the objective becomes more and more fuzzy - the number, type and intensity of conflicts will grow (Hauschildt 1977, p. 215 ff.). There are two forms of complexity which may be mutually reinforcing:

- *Problem complexity*, determined by the characteristics of the actual innovation problem to be solved.

- *System complexity*, resulting from the characteristics of the firm in which the innovation process takes place.

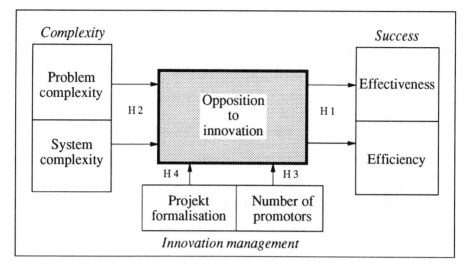

Figure 1: Opposition to innovations – the model

We assume the following relationship between complexity and resistance:

HYPOTHESIS 2:
As the complexity of the problem and of the system increase, so does opposition to the innovation.

The management of innovations is by no means totally at the mercy of these influences. It does have a well-stocked and well proven toolkit at its disposal with which it can proactively influence effectiveness and efficiency.[2]

We consider two of these tools in this study:

(1) The *involvement of promotors* has a positive impact on the success of an innovation. Promotors who support an innovation process actively and intensively (Witte, 1973, p. 15 ff.) are able to re-

---

[2] See, for example, Albach (1993/94, p. 27 ff., 491 ff.), Brockhoff (1994, p. 111 ff.), Specht/Beckmann (1996, p. 114 ff.), Bürgel et al. (1996, p. 151 ff.), Hauschildt (1997, p. 151 ff.).

solve conflicts to the benefit of the innovation project. It is assumed that the promotors themselves co-operate without conflict:[3]

- The *power promotor* overcomes conflicts of power, departmental responsibility and distribution,
- The *technology promotor* resolves conflicts of know-how and perception,
- A third promotor - the *process promotor* – acts as coordinator between the parties affected by and engaged in the conflict.

Accordingly, the third hypothesis is:

HYPOTHESIS 3:
As the number of promotors increases, the opposition to innovation decreases.

This hypothesis assumes *proactive* innovation management, i.e. promotors become involved when they expect resistance or when they receive weak signals of opposition, in order to counter the possibility of prevention, change or delay in the innovation at its inception.

However, they will not behave in this way automatically. It is also possible that promotors will only go into action when huge resistance becomes manifest. In this case, innovation management is *reactive*. Hypothesis 3 would have to read: as opposition increases, the number of promotors countering it will rise.

Because of basic potential validity problems, we did not attempt to cover the contrasting constellations in our questionnaire survey. We therefore have to let the data decide whether proactive or reactive innovation management has been chosen. If there are no findings, both patterns of behavior may be equally probable.

(2) Irrespective of the existence of promotors, the literature on project management leaves us in no doubt that *institutionalization of the innovation as a formal "project"* is an essential prerequisite for project success. It appears that such proactive formalization attracts a higher level of attention, makes it possible for the project to be more clearly separated from operational activities, ensures the provision of resources and facilitates access to the formal techniques of

---

[3] See the summary of the research into the promotor concept in Hauschildt/ Gemünden (1998).

project planning. However, empirical tests of this assertion have not removed all doubts of the effectiveness of such measures in innovative projects[4]. We therefore intend to test it again:

HYPOTHESIS 4:
As the degree of formalization of the innovation as a project increases, opposition to the innovation becomes weaker.

The above considerations also apply accordingly to reactive innovation management: it is possible that manifest resistance forces the innovators to formalize projects.

2. Design of the empirical study

The following empirical research is based on the applications for the "*1998 German Industry Award for Innovation*" which is presented every year by "Wirtschaftswoche" (weekly business journal) in collaboration with the "Wirtschaftsclub Rhein-Main" every year. 1998 was its eighteenth year.[5] 305 companies applied, with 343 innovations. This statistical population represents *clearly defined innovations*, so that questions about attributes are unproblematic. Moreover, it corresponds to managers' understanding of innovation. The disadvantage proved to be its *heterogeneity*: the innovations display extremely wide variations in the level of technological sophistication, and also accommodate innovative applications. For example, the innovative integration of the dynamo, the starter and the active vibration absorber for cars was equivalent - in the subjective opinion of the applicants for the award - to an innovative way of marking out graves. Finally, the variance in the assessments of success is likely to be smaller than in a representative sample, since the applicants are convinced that their innovations are successful. To this extent, there is a "*positive bias*".

On the date of publication of the above issue of "Wirtschaftswoche", the applicants were sent a *three-page questionnaire*. This was

---

[4] See, for example, Rubenstein et al. (1976, p. 17 ff.), Ettlie (1983, p. 237), Ebadi/Utterback (1984, p. 580).

[5] The award winners, the panel of judges (p. 74) and all applicants with their innovations (p. 58 ff.) were named in Wirtschaftswoche No. 4 of 15.1.98.

followed up by telephone calls after four weeks. The responses come from 161 firms and cover 164 innovation projects. In 27 cases, the applicants explicitly refused to respond to the questionnaire. The *return rate* was good, at 52.8% of companies and 47.5% of projects.

Following elimination of the inconsistent responses, the analysis was based on a *sample* of 154 innovation projects from 151 firms.

In the final analysis, a questionnaire survey can only capture the "rational surface" of opposition which we discussed above. It would only be possible to determine the underlying reasons by in-depth interviews or experiments, if at all. However, the questionnaire method was adequate for our purpose of getting started in the field of research into resistance.

3. The characteristics of opposition

3.1 Opposition as a fact

In view of the well known tendency of business practitioners to avoid conflict, it seemed to us that introducing the terms "resistance" or "opposition" without preparatory questions would pose problems. We deliberately did not distinguish between resistance and opposition here. The first question asked was therefore whether the innovation "had always met with the unreserved approval of those affected and those involved" within the company. The answer was "yes" in 63% of cases, in 37% there had been "doubts and objections". This in itself is a finding: *innovation is by no means always accompanied by opposition*. Some of the respondents wrote separate letters on this subject, complaining that we were quite clearly taking an overly negative view of reality. Nevertheless, the positive bias mentioned above has to be taken into account in the quantitative analysis.

3.2 The arguments of opposition

In order to capture the substance of the arguments used in opposition, we gave the respondents a *list of 20 standard responses* and a free space for arguments which had not been included. Multiple mentions were possible. In formulating the questions, an attempt

1. Technical arguments: (78 mentions; 26.9%)

– The new product or process does not work technically,

– maybe a prototype might work at best (but that doesn't mean that the final product or process will work later as well),

– the technical environment is not ready for the innovation,

– the company does not have the right staff to implement the innovation.

2. Market-specific arguments: (84 mentions; 29.0%)

– There isn't enough demand for the innovation,

– it will be impossible to find the right alliance partners.

3. Financial and commercial arguments: (65 mentions; 22.4%)

– The innovation is not financially viable,

– is too expensive,

– cannot be financed,

– destroys economic value.

4. Legal and contractual arguments: (34 mentions; 11.7%)

– There are legal (patent-related) or official objections to the innovation,

– it cannot be protected,

– it will just be imitated.

5. Vague arguments with no specific direction:
(29 mentions; 10.0%)

– The innovation is before its time - or after its time,

– is too risky,

– could have unknown ecological effects,

– threatens jobs,

– no-one is responsible for it,

– and anyway: the status quo is really not at all bad.

Table 1: The arguments used to resist innovations

was made to use ordinary business language, not that of management theory. The questions were directed at several groups of responses which, however, were not presented en bloc in the questionnaire. Table 1 shows how the individual groups of arguments were worded and their absolute and relative frequencies.

Technical and economic arguments are by far the most popular. It seems worth noting that the "vague" arguments very rarely appear. The free space offered was not used either. At the same time, these observations validate the survey. Overall, the picture is as expected, even though we were not able to specify it beforehand in a directed hypothesis. *Opposition to an innovation, like its success, lies in the realm of technical, market-specific and commercial aspects.*

3.3 The intensity of opposition

The content of these opposing arguments was not specifically considered in the further analysis. What we wanted was to derive from the arguments a new measurement of the *intensity of opposition* ("breadth of opposition"). This breadth of opposition takes into account the fact that the quality and quantity of opposition will change if arguments from different functional departments or argument groups are used.

The breadth of opposition is defined by the following conventions and is split up as shown in Table 2.

A further question focused on whether the opponents were prepared to present their arguments *openly*, or whether they preferred to play their cards close to their chests. In those cases where opposition occurred at all, the majority was *covert* (53%), in the remaining 47% opponents expressed their arguments openly. This readiness to come out into the open is an important indicator of the *culture* in which conflicts are handled. Overt opponents display self-confidence and commit to clear responsibility.

1 = No opposing arguments
  (25 mentions; 22.7%)

2 = Only one opposing argument, i.e., for example, only technical arguments or only market-specific ones or only financial/commercial arguments
  (23 mentions; 14.9%)

3 = Opposing arguments from two areas, i.e., for example, technical + market-specific arguments or market-specific + financial and commercial arguments or technical + financial/commercial arguments
  (47 mentions; 30.6%)

4 = Opposing arguments from three (and more) areas, i.e., for example, technical + market-specific + financial/commercial arguments
  (49 mentions; 31.8%)

Table 2: The breadth of opposing arguments

3.4 The assessment of opposition

The respondents were asked to assess the resistance. These assessments naturally included the observed effects of opposition. It therefore seemed difficult to make a clear distinction between form and primary effects, and assessment of the effects was therefore integrated into the assessment of opposition. The questions and answers in this area were as follows:

- Did opposition *delay* or *change* the result?
  Delay was identified in 12% of the projects, change in 4.7%, both effects occurred in a further 4.7% of the projects.

- With hindsight, would you assess the effect of opposition as *harmful, useful or insignificant* ?
  In 12% of the projects opposition was characterized as harmful, in 11.3% as useful and in 20% as insignificant.

- Finally, we asked whether the project had been *threatened with cancellation* one or more times during its implementation.
  This had happened at least once in 16.7% of cases, several times in a further 12.7% - a clear sign of the effectiveness of opposition.

3.5 The variants of resistance: "destructive" and "constructive" opposition

In accordance with the heuristic mission of this study, the individual results reported above were further condensed by factor analysis. Table 3 shows the result.

	Factor 1	Factor 2	MSA values
Covert opposition	**.83810**	-.07747	0.53
Breadth of opposition	**.70867**	.07096	0.79
Project cancellation	**.60784**	.21691	0.73
Intention: delay	**.57513**	.38233	0.71
Overt opposition	-.02766	**.83450**	0.48
Assessment "useful"	.41766	**.69697**	0.60
Intention: change	.08704	**.65868**	0.66
Explanation of variance (in %)	37.6	18.3	
Cronbach's alpha	0.67	0.65	

The factors were extracted after the analysis of the main components.
Orthagonal rotation was carried out in accordance with the Varimax method.
The KMO value is 0.62. The Bartlett test of sphericity is highly significant ($p < 0.000$).
The MSA values are over 0.5 with one exception.

Table 3: Typology of opposition (factor analysis)

The two factors are termed "*destructive*" (Factor 1) and "*constructive*" *opposition* (Factor 2), in accordance with Klöter's suggestion (1997, p. 161).

- *Destructive opposition* remains covert. The breadth of opposition is greater, i.e., it uses several opposing arguments of different types. It is intent on having the project cancelled, i.e., it wants to prevent the innovation entirely. It aims at least to delay the project.

- *Constructive opposition* expresses its concerns overtly. It is deemed to be "useful". It aims for modification of the innovation.

The ambivalent assessment of opposition presented above is thus quite correct. It cannot and should not be assumed that the conflicts which come to light in the opposing arguments are always assessed as harmful.

These existential conclusions thus show variance, and indicate that it is justifiable to test the hypotheses on the causes and effects of opposition.[6]

4. Opposition to innovation - its context and its results

Hypothesis 1 contends that there is a negative correlation between resistance on the one hand and effectiveness and efficiency on the other. The measures of success were operationalized as follows:

– *Effectiveness*: assessment of the *degree of innovativeness* of the innovation on a scale of three with the following graduations: "technically completely new solution to a known application" (24.2%), "development of new applications" (27.5%) and "both" (48.3%).
  In addition, the "*economic success of the project*" compared to expectations was asked for and evaluated on a scale of 7. The mean (M) is 3.98, the standard deviation (SD) is 1.55, the ratings show a good normal distribution.

– *Efficiency*: assessment (on a scale of 7) of the "*costs* of the project compared to their original estimates" (M = 3.66, SD = 1.26) and the "realization of *time* objectives" (M = 3.23, SD = 1.30), both with a good normal distribution.

The remaining hypotheses refer to influences of problem and system complexity and of the tools of innovation management on resistance. All hypotheses can be tested in aggregate in a path analysis. This has the added advantage of demonstrating those relationships between the context and success which have nothing to do with opposition.

Figure 2 shows the relationships with success in the path diagram and admits the following interpretation:

---

[6] We use factor values below as variables, well aware of the fact that the objection of overfitting will be raised. However, the number of variables involved is small and controllable. The factor weightings of the items are not far apart. And finally: the characterization and naming of the factors is based on theoretical prior thinking.

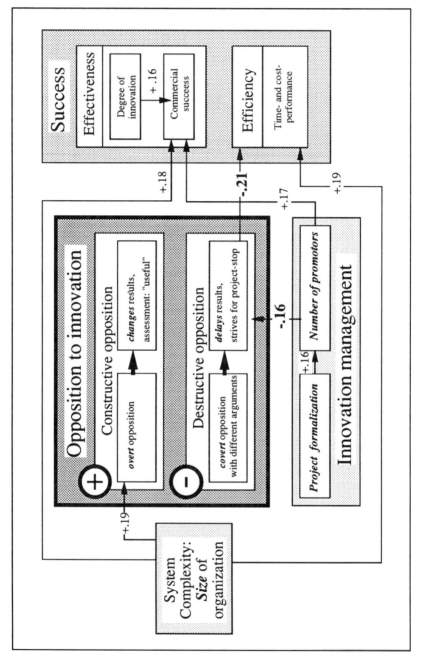

Figure 2: Opposition to innovations and its results

4.1 The relationship between opposition and success

There is no positive correlation between *constructive opposition* and success. The positive assessments of opposition cannot be directly proved, at all events. At best, the old German saying that "success has many fathers" is confirmed - evidently including the constructive opponents.

Destructive opposition is obviously effective in its attempts to delay innovation projects, at least, the efficiency of the projects is adversely affected. This is at the same time a sound validation of the assessment of the primary effect. H1 cannot be rejected for this variant of opposition.

4.2 The relationship between problem complexity, opposition and success

Problem complexity was operationalized using the following variables:

- Project *duration* (in months): M = 23.57, SD = 21.35,
- Project *expenditure* (in % of annual turnover): M = 20.6, SD = 29.67,
- Project *staff* (capita): M = 5.74, SD = 4.93,
- Project *competition* (number of simultaneously conducted competing projects): M = 6.8, SD = 42.9.

To our surprise, none of these variables show a significant correlation with opposition and with the various measures of success. Evidently, the projects can be classed as having the same or at least similar task complexity[6], so that the differences found in our sample are irrelevant. H2 is definitely rejected with regard to problem complexity.

---

[6] See Lechler (1997, p. 238 ff.).

4.3 The relationship between system complexity, opposition and success

We operationalized *system complexity* by *size* of firm (in terms of annual turnover) and by *industry*. The industry is a measurement of doubtful value, everyone uses it and it rarely shows clear results. It certainly did not do so in this case. In contrast, the relationship between size of firm and the observed phenomena did produce results: not only do direct correlations with effectiveness and efficiency exist, but an indirect one with constructive opposition was also identified.

A close examination of the data reveals a *skewed size distribution*. In our sample, the *mean* of annual turnover for 1997 was DM 2.8 million, with a standard deviation of DM 12,015 million! In other words, very small "inventor-entrepreneurs" are competing with the innovation teams of medium-sized firms and the R&D departments of large ones. In order to take these differences into account robustly, we replaced size of company in the path diagram by a 0/1 variable which split out all firms with an annual turnover of less than one million marks (0) from all other categories (1). Finer size distinctions than that do not affect the result.

If this simplification is accepted, the following conclusions can be drawn: *inventor-entrepreneurs* do not practise any division of labor when identifying and realizing the innovation. They therefore *encounter no opposition*, or are able to suppress it at its inception due to their position of superiority. The frequently-cited entrepreneurship of new business start-ups and of personally responsible, risk-taking entrepreneurs is demonstrated here.

There is a contrast here with the *large firms*, where innovation is a process conducted by division of labor and has to integrate a large number of individuals who are either actively involved or passively affected. As the findings demonstrate, opponents can definitely be won over to constructive collaboration. To this extent, H2 is rejected for system complexity as well.

The direct correlation between size of firm and success is clearly attributable to the specific features of our sample. In larger firms, there is likely to be more than one innovation. They can therefore present their company's best solution in the application for the innovation award. Projects are thus pre-selected on success-related grounds.

4.4 The relationship between innovation management, opposition and success

Innovation management was reduced to the influence of promotors and project formalization.

We deliberately avoided the suggestive term *"promotors"*, and instead asked for the following:

- "ongoing *support from top management*" (we meant by this the influence of a power promotor),

- the existence of a *"particular expert* who gave special technical backing to the project" (by which we meant the technology promotor),

- the existence of a "dedicated *project manager* in addition to the technical expert" (meaning the process promotor) and

- "other people who gave particular commitment to the project".

Deliberate innovation management by committed promotors overcomes destructive opposition; or, more precisely: *the number of promotors is important*. The three-promotor constellation not only occurs more frequently (in 95 cases) than the two-promotor structure (28 cases) and the single promotor (11 cases). The three-promotor constellation also has the highest share of solutions classed as above average and the lowest share of those classed as below average in terms of economic success.

However, the promotors do not only have a direct influence on effectiveness. Above all, they are in a position to actively counter destructive opposition. H3 is not rejected; indeed, the influence of the promotors goes far beyond the management of conflicts with opponents.

As a result, the importance of *proactive versus reactive innovation* management can also be assessed: the firms had behaved in a proactive manner to a significant extent. However, the findings are not overwhelmingly high either. It is quite clear that further studies are needed to identify the different effects of proactive and reactive innovation management.

The formalization of the innovation as a *"project"* does not accomplish anything comparable to the activity of promotors. On the face of it, H4 is rejected. Yet formalization of the project does make an important contribution, since it clearly fosters the development of promotor structures. This finding demands further research into possible relationships between the techniques of project management and the promotor concept.

5. Discussion

5.1 The results

In view of the intuitively convincing discussion in the literature *the* negative effects of opposition, we were surprised to find that *absolutely no opposition* was registered in a large number of the innovation projects studied here. This is largely due to the fact that in more than 20% of our cases the innovations were generated by inventor-entrepreneurs, where the only conflicts likely to occur are interpersonal ones which are difficult to identify in a questionnaire survey.

When resistance occurs in larger firms with a higher degree of division of labor, its impact is *not necessarily negative*. The most important finding *is* "constructive opposition", which concentrates on one or on a small number of criticisms, argues openly and aims to change the innovation, or, more precisely, to improve it. This behavior is also positively assessed by the promotors of the innovation.

Nevertheless, "destructive opposition" occurs in many cases. This type of opposition disguises itself, bases its attacks on a very wide range of arguments, changes tack when the original arguments are refuted, is totally persistent in pursuing its goal of preventing or at least delaying the innovation. It seems that only committed promotors who are superior to the opponents in technical knowledge, diplomatic skills and hierarchical potential can counter destructive opposition.

In addition to these findings, a number of *non-findings* appear to us to be worth mentioning:

– Opposition appears to be independent of the formalization, duration, size and competitive situation of the innovation projects. Resistance evidently arises to the fact and the content of the innovation, and less so to project-specific characteristics. Problem complexity as such does not appear to affect opposition in this study.

- The content-based arguments are independent of the fact, the intensity and the assessment of the opposition. It cannot be said, for example, that constructive resistance is mainly based on technical arguments and destructive resistance on economic ones.

- Finally, it was examined whether opposition decreases if the initiative to generate the innovation comes from members of top management or from the customer. Hierarchical pressure or market pull could be expected to make opposition hopeless from the start. Our material did not support this notion either. Opponents are evidently not intimidated by these variants of compulsion.

We have had to point out at various times that the sample used here determined the results. To this extent, it is impossible to generalize from our findings at this stage. In view of the high practical significance of opposition to innovations it seems to be definitely worth while to pursue further research into the issues raised here.

5.2 Possible implications

With all due caution, however, the following thoughts on practical handling of opposition to innovation do appear reasonable:

The familiar methods for *overcoming destructive opposition* can and should be maintained and further developed. That applies in particular to further efforts to extend the promotor concept and to adapt it to new forms of resistance. In addition, the aspect of opposition should be added to the normative, generally practice-based concepts of project management.

The deliberate *encouragement of constructive opposition* emerges as a new task of innovation management. Markham et al. (1991, p. 237) make the same demand: "Although the antagonist role may be seen in a negative light, we suggest that management consider the antagonist an ally in fostering constructive opposition."

First of all, it seems that practitioners still have to be told that conflicts and differences of opinion should not be regarded as always negative, but may very well be the source of creativity and improvement. Our findings give a first indication of how constructive opposition can be identified: those who strive for change, and if possible for improvement of the innovation, i.e. do not reject it from the start, those who focus on a key criticism - however "irritating" they may be - deserve to be involved in the innovation process and not to be opposed themselves.

Those, on the other hand, who do not appear on the scene overtly as opponents, but as fault-finders who continually advance new arguments to back up their criticism - who may be "false friends" in disguise - who delay the innovation continually by new discussions of first principles, by asking for yet more expert reports, endless consultancy, changing coalitions, are open to the suspicion that their ultimate goal is to prevent it. No constructive opponent does these things.

It is therefore crucial to recognize the direction of thrust of the opposition early, to take prompt action to integrate the constructive opposition and prevent destructive opposition from developing.

This raises the final question: can destructive opposition be transformed into constructive opposition? The remit of management science ends here, and that of psychology and education begins. However, we are more inclined to be skeptical than hopeful.

Starting conditions of successful European R&D consortia

**Hans Georg Gemünden /
Martin Högl / Thomas Lechler / Alexandre Saad**

Starting conditions of successful European R&D consortia

Hans Georg Gemünden /
Martin Högl / Thomas Lechler / Alexandre Saad

1. Introduction

The European Union spends large amounts of money to fund such European research programs like BRITE EURAM or ESPRIT. In these programs corporations and universities from different European countries form R&D consortia in order to cooperatively conduct research and to develop and apply new technologies. Through these programs the European Union hopes to bundle the competencies of its member countries and to develop and exploit innovative solutions based on leading-edge technologies in order to increase the competitiveness, employment and wealth of its economies.

However, much criticism has been raised challenging the efficiency and effectiveness of such programs. The critics raise objections against inefficient bureaucratic procedures, high expenditures, and uncertainty of public sponsorship, which prevent particularly small and medium-sized innovators from participating in such programs. They criticize political restrictions and influences, e.g., aims to fulfill national quota and to include members from less-developed European countries. They object that really promising important R&D projects would have also been performed without European funding, and they assert that the technological know-how created through the programs is not exploited enough.

It is very difficult to examine to what extent these criticisms are justified. The projects which are funded concentrate on new technologies and complex tasks. These projects are characterized by high technical risks. The cooperation of partners from different countries with different cultures, competencies, roles and interests breeds major social and management risks. Furthermore, the outputs of the projects are often intangible assets, knowledge gain, improved standards, prototypes, i. e., at best embryonic solutions far away from large-scale commercial application and thus still containing considerable exploitation barriers and market risks. Hence, there are only limited possibilities to assess the economic potentials of the funded projects, and it is hard to assess the technical risks.

Despite these difficulties, which are inherent in innovative projects, we have the impression that such programs are important to facilitate exchange and cooperation between institutions (firms, universities, and other research institutions) from different countries. Furthermore, we believe that many of the funded pre-competitive cooperative projects would not have been undertaken without this public funding. Several large-scale empirical investigations show that firms which are actively engaging in R&D cooperation and which have more elaborated technology networks also realize higher levels of process and product innovation success.[1]

Thus, in principle, the programs make sense, particularly if compared with the expenditures of the European Union for agriculture or for low-tech industries. The programs are important to build up new industries and to support the establishment of open technical standards. But we need better tools to plan, select, and control such programs and funded projects. Such tools should be developed on the basis of empirical knowledge.[2]

The unit of analysis of this investigation is the *project*, not the program. Assuming that a program like ESPRIT III is economically and politically justified, we want to identify the characteristics of

---

[1] For analyses of the relationship between technology networks and innovation success: Håkansson 1987, 1989; Gemünden, Heydebreck, Herden 1992; Gemünden, Heydebreck 1994; Wolff et al. 1994; Gemünden, Ritter, Heydebreck 1996; Heydebreck 1996; Amelingmeyer, Gerhard, Specht 1997; Gemünden, Heydebreck 1997; Gemünden, Ritter 1998a; Gemünden, Ritter 1998b; Ritter 1998.

[2] For empirical studies assessing the impact of R&D cooperation with other firms on innovation success: Håkansson 1987, 1989; Roos 1989; Fornell, Lorange, Roos 1990; Rotering 1990; Brockhoff, Gupta, Rotering 1991; Brockhoff 1992; Gemünden, Heydebreck, Herden 1992; Niosi, Bergeron 1992; Håkanson 1993; König, Licht, Staat 1993; Linné 1993; Becker 1994; Hagedoorn, Schakenraad 1994; Kreiner 1994; Teichert 1994; Wolff et al. 1994; Waudig 1994; Aldrich, Sasaki 1995; Becker 1995; Bruce et al. 1995; Barker, Dale, Georghiou 1996; Heydebreck 1996; Tucci 1996; Olk, Young 1997; Kropeit 1998; Ritter 1998; and Saad 1998 with further references.

successful projects. More specifically, we want to identify the *critical starting conditions*, i. e., the critical success factors that should be given at the *start* of a publicly funded European cooperative R&D project. If a model is able to sufficiently explain relevant criteria of project success with a small set of generally important starting conditions, then it can be used to improve the formation and selection of R&D consortia. There are several institutions and people that may substantially benefit from the results of this empirical investigation: Participants of cooperative R&D projects, political decision makers, reviewers involved in the selection process, and consultants helping the participants to form their consortia and develop their proposals.

1. The *participants* can find out if their project poses risks that they have not yet identified. According to our findings, each of the typical seven or eight participants of an ESPRIT project invests three to four person months into the development of a proposal. They can plan and perform actions to reduce these risks or to revise their decision for the project if necessary.

2. The *political decision makers* and *reviewers* who are responsible for the evaluation and selection of the projects can revise their set of criteria and use our reliable and valid diagnostic instruments to improve the selection. They can also formulate additional requirements, which the participants should fulfill in order to receive this public funding.

3. The *project consultants*, employed in the different countries to support the formation of R&D consortia and the formulation of proposals, can use our findings to inform their clients which criteria their partners should fulfill and which requirements the interorganizational cooperation must satisfy.

Combining the effects reached through all three classes of participants it is possible to improve the efficiency and effectiveness of the programs. It would mean an increase of effectiveness of 50%, if the rate of really successful projects could be raised from 20% to 30%.

2. The theoretical framework

The basic hypothesis of the investigation is: *The better the starting conditions of a cooperative R&D project, the better the progress of this project and the higher the success of the project.*

But, which starting conditions are relevant? The very extensive literature on cooperation, strategic alliances, joint ventures, networks in general, and the specific literature on bilateral and multilateral R&D cooperations offers long lists of desirable characteristics, which the partners of a cooperation should fulfill. A multitude of ethical, cultural, strategic, organizational, size and task-related "fits" should be given in order to make a cooperation a successful one (Kirchmann 1994, Raffée, Eisele 1994;Harvey, Lusch 1995; Lubritz 1996; Kropeit 1998; Saad 1998).

The main objective of this article is to derive a core model of starting conditions critical to the progress and the success of European cooperative R&D projects. This core model is based on three kinds of fits: *resource fit, social fit and goal fit.*

In the process of forming the R&D consortium the partners have to find other partners with whom to conduct the project. Chiefly the content, the quality, and the compatibility of resources contributed by the partners govern this choice of partners. Thus, partner search and partner choice dictates the potential competencies and synergies of the consortium.

A good resource base is a necessary but not a sufficient condition for the success of a cooperation. As the partners must be able to work together, they must establish a social fit. Mutual trust and commitment to the joint project must be established in order to achieve high quality teamwork within the consortium (For the influence of teamwork on innovative projects and on cooperations see: Scott 1997; Gemünden, Högl 1998; Helfert 1998; Högl 1998; Högl, Gemünden 1998a, 1998b, 1999 with further references).

In this research, the unit of analysis is the project. The empirical studies on success factors of project management strongly suggest that sufficiently clear goals are needed in order to direct the project, to motivate participants, and to coordinate different parties as well as their contributions. Even termination of a project is very difficult if the goals are unspecific and unclear. Furthermore, as the projects investigated in this research are being conducted cooperatively by several independent parties (i.e., the consortium), it is of crucial importance that the members of the consortium have common interests and thus seek a common goal. A lack of commonality poses the inevitable risk of severe goal conflicts during the project, which, in turn, is detrimental to the ultimate success of that joint endeavor.

Let's take a closer look at theses three fits and their six components in order to derive our basic model.

2.1 Resource fit: Competence and synergy

The better the technological competencies suit the requirements of the task, and the better the competencies of the partners complement each other, the higher are the chances for a successful cooperative project. The competencies of the consortium refer to the technological position of the partners. The technological position relates to factors such as knowledge, skills, and rights, as well as achieved reputation and resources creating results such as products, processes, patents, publications, and networks. These are positive conditions for top performance. A number of studies show that high competencies are important for the success of R&D cooperations (Björkman et al. 1991; Brockhoff, Gupta, Rotering 1991; Wolff et al. 1994; Niosi, Bergeron 1992; Sakakibara 1993; Lee, Lee 1993; Lee, Lee, Bobe 1993; Bruce et al. 1995). Additionally, it is generally the case that strong partners tend to cooperate with their like. Hence we formulate the following hypothesis:

HYPOTHESIS 1:
The greater the consortium's competencies at the beginning of a cooperative R&D project, the greater the project success will be.

Why should highly competent partners work with one another, as they will have to share the yields of their work with one another accordingly? The answer is that they will prefer to cooperate with other competent partners if the project cannot be successfully tackled alone because each single partner lacks critical resources. There are several possibilities for such synergies:

(1) Partners from different stages in the value chain (producers and users; scientists and industry; etc.) combine their competencies.

(2) Partners from different disciplines contribute to a joint project.

(3) Partners from different countries and cultures cooperate.

These types of alliances enter new directions of thinking and thus create new concepts, prototypes, and products featuring entirely new functionalities. It offers a particularly high and challenging potential to learn from one another and to explore novel solutions (Doz 1988; Souder, Nassar 1990; Dickson, Smith, Smith 1991; Linné et al. 1991; Sinha, Cusumano 1991; Dobberstein 1992; Niosi, Bergeron 1992; Hagedoorn 1993; Ormala et al. 1993; Teichert 1994; Barker, Dale, Georghiou 1996). This results in the following hypothesis:

HYPOTHESIS 2:
The greater the complementarity of the technological resources at the beginning of a cooperative R&D project, the greater the project success will be.

Authors often recognize complementarity of resources as the main prerequisite for cooperation. It is also often named the main motive for cooperating. While a number of authors explain that the partners of a consortium are heterogeneous and want to stay that way and will stay that way, others point out that cooperating involves sharing of know-how and thus poses the threat of losing the exclusive possession of a crucial resource. In that sense, Japanese-US alliances have often been described as a race to see who can learn the most from each other first. The danger of taking advantage of the other partners is particularly severe if the partners operate at different levels of power, if the partners' dependence on the project strongly differs, and if it is not a symmetric relationship. Competencies and complementarity, therefore, are necessary prerequisites of a successful cooperation; however, they do not guarantee success. Other conditions, such as a high level of social fit have to be met as well.

2.2 Social fit: Trust and commitment

Successful cooperations require not only highly competent and compatible partners, but it has to be ensured that the potential of each partner is fully utilized through the cooperative work in the consortium. This demands *trust* among the partners that other partners will not exploit their vulnerable know-how or transfer it to third parties, without the prior consent of that partner. Hence, trust is a necessity to actively engage in a cooperation. The term *trust of the consortium* refers to the willingness of partners of a European cooperative R&D project to take on risks in light of certain expectations. This includes the optimistic assumption, that the other partners can be trusted as they are perceived as open, honest, reliable, and benevolent (Bachofner 1996; Saad 1998; Butler 1991; Loose, Sydow 1994; Morgan, Hunt 1994; Wurche 1994; Mayer, Davis, Schoorman 1995; Plötner 1995; Fennetau, Guibert 1997; Smith, Barclay 1997).

The trust of the consortium creates an environment for open communication early in the project, prevents restrictive bureaucratic rules, and creates flexibility. Someone who is trusting is already prepared to sacrifice control and is also prepared to contribute to the project whatever it takes, expecting the partners to do so as well. For these reasons, the effort exerted on controls is reduced, and processes of learning are accelerated and improved (Hull, Slowinsky 1990; Linné et al. 1993; Håkanson 1993; Ormala 1993; Inkpen, Birkenshaw 1994; Kreiner 1994a, 1994b; Bruce et al. 1995; Eisele 1995; Fontanari 1996; Kemp, Ghauri 1997). Therefore we postulate the following hypothesis:

HYPOTHESIS 3:
The greater the trust of the consortium at the beginning of a cooperative R&D project, the greater the project success will be.

A second aspect of the 'social fit' is the partners' commitment with regards to the consortium. The term commitment to the consortium refers to the willingness of the partners of a European cooperative R&D project to enter obligations and to live up to them. This self-obligating behavior results from a long-term perspective and an orientation on common goals. This is expressed by the willingness to invest and sacrifice, even if short-term disadvantages are incurred (Söllner 1993; Morgan, Hunt 1994; Bachofner 1996; Saad 1998). Such commitment is necessary to actually build a consortium and formulate a common proposal. Without a certain level of commitment, the partners will not invest their time or money to develop an ambitious, well-coordinated and supportable proposal.

Commitment is a necessary requirement, but does not ensure project success. An intensive inquiry by Kreiner (1994a, 1994b) into EUREKA Consortia involving Danish partners documents that such projects are characterized by highly turbulent environments. This is why a number of consortia ended in partners leaving and the ultimate failure of the projects. High commitment is grounds for accepting common norms, procedures, and interfaces. Thus, it creates stability, improves the collaboration and helps the consortium to deal with intercultural differences. High commitment reduces the propensity to leave the consortium and, in turn, increases the tendency to adapt to environmental change with a collective search for new solutions (Roos et al. 1989; Hull, Slowinsky 1990; Souder, Nassar 1990; Björkman et al. 1991; Dickson, Smith, Smith 1991; Wissema, Euser 1991; Farr, Fischer 1992; Kreiner 1994a, 1994b; Bruce et al. 1995).

A high level of commitment should lead to higher levels of group cohesion and identification with the group. This is why the explanations and the empirical evidence regarding the relationship between group cohesion and group performance, particularly with innovation teams, are relevant to the issue at hand (Scott 1997; Högl, Gemünden 1998a, 1998b, 1999). Hence we postulate the following hypothesis:

HYPOTHESIS 4:
The greater the commitment at the beginning of a cooperative R&D project, the greater the project success will be.

2.3 Goal fit: Clarity and compatibility

Resource fit and social fit ensure that the consortium embodies a high level of compatible competencies and that it actively utilizes this potential. However, a third necessity for project success is still missing: The resources must be employed to achieve a common goal. Therefore, a common direction has to be established.

This complementarity of goals, called goal fit, includes two dimensions, goal clarity and goal compatibility. Goal clarity refers to a goal system being understandable, measurable, controllable, and realistic. It is important that clear personal and/or institutional responsibilities are associated with the goals and their attainment (Gemünden 1995; Hauschildt 1997). Goal clarity has often been shown to be an important predictor of project success (Gemünden, Lechler 1997; Lechler 1997). If the project is to progress well, then clear goals have to be established in the early phases of the project, so that contributions from the members of the consortium can be defined and coordinated during the course of the project. Consequently, we postulate the following hypothesis:

HYPOTHESIS 5:
The greater the goal clarity at the beginning of a cooperative R&D project, the greater the project success will be.

High goal clarity not only depends on the mental penetration and structurization of the problem at hand, but also requires common and compatible perceptions of the partners. If the participants of the consortium seek after different goals, then they will likely refuse to commit themselves to certain goals, in order not to question the overall project. If, however, great conflicts of interests are present that can only be resolved by pseudo consensus, then changes of goals later in the project are inevitable. Therefore, we expect that the synergies created by compatible resources can be exploited only, if the partners' interests are compatible (Doz 1988; Nueno, Oosterveld 1988; Roos 1989; Evan, Olk 1990; Wissema, Euser 1990; Farr, Fischer 1992; Lee, Lee 1993; Linné et al. 1991; Sakakibara 1993; Bruce et al. 1995; Eisele 1995). Hence, we postulate the following hypothesis:

HYPOTHESIS 6:
The greater the compatibility of the goals at the beginning of a R&D project, the greater the project success will be.

2.4 Hypotheses regarding the relationships between the starting conditions

A powerful explanation of the effects of the starting conditions can be achieved when also considering causal relationships between them. Then path analysis can be employed to estimate direct and indirect effects of the starting conditions on project success.

So far, we have described six starting conditions, so that we have to derive $6*5/2 = 15$ hypotheses regarding causal relationships. This, however, makes little sense, as we would need an unrealistically large sample in order to test such a model. Therefore, we restrict ourselves to, in our view, the most important starting conditions, namely the competence, the trust, and the clarity of the goals. In doing so, we pick one factor each from the areas of resource fit, social fit, and goal fit, and thus touch upon all the main areas of the base model.

We assume that high competence represents an important requirement for high levels of trust in a consortium. If a partner is regarded asless competent, then one is less likely to trust this partner to live up to expected performance. Empirical investigations regarding the causes of trust in business relationships, cooperations, and international joint ventures describe perceived competence of a partner to perform at expected levels to be an integral part of the construct trust (Morgan, Hunt 1994; Plötner 1995; Fennetau, Guibert 1997; Smith, Barclay 1997). Consequently, we formulate the following hypothesis:

HYPOTHESIS 7:
The greater the competence of the consortium at the beginning of a cooperative R&D project, the greater the trust of the consortium at the beginning of a cooperative R&D project.

High technical competence is also a prime condition in order to develop realistic and clear goals. In other words, if someone is unfamiliar with the innovative matter at hand, then this person will be unable to arrive at clear goals. Specifying goals, such as a list of requirements for a prototype of a product, demands expert knowledge of the field of application. We propose the following hypothesis:

HYPOTHESIS 8:
The greater the competence of the consortium at the beginning of a cooperative R&D project, the greater the goal clarity at the beginning of a cooperative R&D project will be.

Clear goals are not only understandable and traceable, but their achievement can also be more easily controlled, and the compliance with predetermined rights and responsibilities is more easily determined. Insofar the willingness increases to trust partners that specify such goals. Hence, we postulate the following hypothesis:

HYPOTHESIS 9:
The greater the goal clarity at the beginning of a cooperative R&D project, the greater the trust of a consortium at the beginning of a cooperative R&D project.

2.5 Hypotheses regarding the relationships between the starting conditions, the project's progress, and project success

If our hypotheses regarding the effects of the starting conditions are true, then the starting conditions should impact not only the (later) project success, but also the project's progress. We are examining the following three variables in this respect:

(1) Goal changes during the project;

(2) Conflicts escalating during the project;

(3) The quality of the project management.

Better starting conditions are likely to result in fewer goal changes, fewer conflicts, and better project management. These variables, in turn, should have a positive effect on the (later) project success. For these reasons, we formulate the hypotheses below:

HYPOTHESIS 11:
The better the starting conditions of a cooperative R&D project, the less likely goal changes are to occur during the project.

HYPOTHESIS 12:
The better the starting conditions of a cooperative R&D project, the less likely conflicts are to escalate during the project.

HYPOTHESIS 13:
The better the starting conditions of a cooperative R&D project, the higher the quality of project management during the project.

We still have no hypotheses regarding the influence of the project's progress on project success in order to complete our path model. Following the literature and empirical evidence from past studies, we expect the quality of project management to have a positive influence, and conflict escalation and goal changes to have a negative influence on project success. Furthermore, we assume that high-quality project management will restrict the likelihood of conflict escalation and goal changes.

HYPOTHESIS 14:
The better the project management of a cooperative R&D project, the greater the project success will be.

HYPOTHESIS 15:
The greater the extent of goal changes during a cooperative R&D project, the smaller the project success will be.

HYPOTHESIS 16:
The greater the extent of conflict escalations during a cooperative R&D project, the smaller the project success will be.

HYPOTHESIS 17:
The better the project management of a cooperative R&D project, the smaller the extent of conflict escalation will be.

Figure 1 summarizes our theoretical framework.

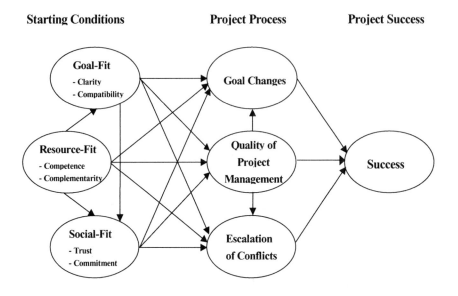

Figure 1: The conceptual framework

HYPOTHESIS 18:
The better the project management of a cooperative R&D project, the smaller the extent of goal changes will be.

3. Methods

3.1 Sample

The present empirical investigation of European cooperative R&D projects was conducted using data from projects sponsored under the ESPRIT III program of the European Commission. A total of 676 questionnaires (including all projects sponsored under ESPRIT III) were sent out in English, French, German, Greek, Italian, and Spanish. Questionnaires were returned by 339 respondents of 10 European countries, resulting in a response rate of 50.1 %, which may be considered as high, given the scale and the complexity of this survey. The 339 responses received refer to 193 ESPRIT projects.

The sample of the surveyed ESPRIT partners with their ESPRIT projects can be characterized as follows:

1. The questionnaires were filled out by the appropriate respondents, underlining the validity of the data gathered:
   - some 70 % of the respondents were the leaders of the project;
   - more than 80 % of the respondents were involved in the project from the beginning;
   - more than 90 % of the respondents were experienced with international R&D projects.

2. The sample includes ESPRIT partners from a wide variety of European countries, where a representative share of all ESPRIT projects could be surveyed;
   - more than one partner was surveyed for about 50 % of the projects;
   - 24 % of the partners and 23 % of the project's budget were surveyed.

3. The sample documents the heterogeneity of the ESPRIT partners and offers a clear picture of the origin, composition, and roles of the partners in the ESPRIT project.
   - ESPRIT partner from 10 European countries;
   - more than 60 % of the partners operate as technology developers;
   - more than 70 % are industrial partners.

4. Even though there are significant differences in the initiating phase of ESPRIT projects (before project start), the average initiating phase of a ESPRIT project can be characterized as follows:
   - Four meetings among the partners.
   - Four person months investment by the leading partner; some three person months for the other partners.
   - Twelve months absolute time consumption for leading partner; some seven months for the other partners.
   - On average, an ESPRIT partner knows two other partners before the project; however, 30 % of the partners do not know any other partner; the share of known partners in the consortium has a median value of 20%.
   - Some 75% of the project proposals are rejected by the European Commission, i. e., they receive no funding.
   - The remaining 25% of the proposals are cut on average by 25% of their proposed budget.

3.2 Measures

3.2.1 Measure of project success

The constructs of our model are latent variables which are not directly observable. They have to be measured by a number of indicators. Project success is a multidimensional construct, hence we operationalize several sub-constructs. Each of them is measured by several items.[3] The level of agreement/disagreement for each statement (item) was rated on a seven point rating scale. The sub-constructs of project success are *overall success, effectiveness, efficiency, collaboration success and learning success.*

The results of our measurement procedures are documented in the following tables. All calculations are based on a sample of 167 projects with nearly complete data. For most of the cases that were excluded the respondents did not answer the questions for the effectiveness indicators. Most of these projects did not result in a prototype which could have been evaluated by the respondent. Including these cases for the other constructs which had no data missing did not change our results. In order to secure comparability of all results we used the same sample of 167 projects for all analyses. Values still missing in that data set were substituted by the mean value of this item. In 87 of the 167 projects we were able to average the answers of two or more respondents from different participants.

We performed reliability analyses to test for convergent validity and homogeneity of the scales. Cronbach's alpha should be higher than .70 and the item-to-total correlations should be at least .30. Factor analysis was used to test whether the items load sufficiently high on only one common factor which explains at least 50% of variance.

---

[3] For similar procedures see: Gemünden 1981, p. 162 ff.; Gemünden 1990; Hauschildt 1991; Kölmel 1996; Hauschildt 1997; p. 389 ff.; Kropeit 1998, p. 187 ff.; Lechler 1997, p. 64 ff., 88 ff.; Lechler, Gemünden 1998; Saad 1998, p. 206 ff.

Overall success: The construct *overall success* is used to attain a general assessment of the success of the cooperative R&D project, including all aspects of success. The developed items intend to measure success from different perspectives of the individual participants, the consortium as a whole and the European commission. A fourth item is added to measure the willingness to participate again in such a project.

Indicator	Item-to-total correlations	Factor loadings
Overall, the project is a big success for your company.	.80	.88
Overall, from today's perspective your company would again participate in this project.	.75	.85
Overall, the project is a big success for the consortium.	.83	.91
Overall, the commission is satisfied by the results of the project.	.73	.84
Summarizing statistics Cronbach's alpha / explained variance	.90	76.0 %

Table 1: Measure of overall success

Effectiveness: The construct *effectiveness* is used to assess the *quality* of the output of a project by comparing the intended output with the actual output. The objective of the ESPRIT projects investigated here is to gain new knowledge, to establish the basis for a new standard, or to develop and test a prototype for a new product. We concentrate in our output assessment on a prototype, but use this term in a broad sense, i.e., it includes not only hardware and software products for end users, but also new programs, devices, and tools to support software development.

The objectives of an ESPRIT project are usually defined in terms of technical specifications, not in terms of commercial goals. Very often grants are given to pre-competitive projects. Although commercial arguments are used to justify the funding of the project, the commercialization of the results achieved during the project is not part of the project itself. This commercialization (if wanted) is usually undertaken after the project and is not necessarily performed collectively and not even by the same members of the funded R&D consortium. Therefore, we restrict our measurement of the effectiveness of the project to the technical success. Measurement of commercial success would have required another research design. We would have had to wait much longer in order to assess the commercial success of the project, and we would have had to consider other factors which also influence the exploitation of the project. Furthermore, it would have been much more difficult to find respondents that participated in the initialization and preparation stage of the project and that were able to remember this stage adequately.

Our items assess to what extent a prototype was implemented, tested, generated satisfactory results for developers and users, and fulfilled the specified requirements.

Indicator	Item-to-total correlations	Factor loadings
The pursued prototype was fully implemented.	.86	.89
The implemented prototype was fully tested.	.86	.90
The evaluation of the prototype generated satisfactory results for the developers.	.82	.88
The evaluation of the prototype generated satisfactory results for the users.	.86	.90
The prototype fulfilled the specified requirements.	.85	.89
Summarizing statistics Cronbach's alpha / explained variance	.94	79.3 %

Table 2: Measure of effectiveness

Efficiency: The construct *efficiency* is used to assess the *quality* of the input of a project by comparing the intended input with the actual input. We use two items which measure the extent to which deadlines and budgets were met.

Indicator	Item-to-total correlations	Factor loadings
All essential project deadlines were met.	.64	.91
All essential budget positions were met.	.64	.91
Summarizing statistics Cronbach's alpha / explained variance	.77	81.9 %

Table 3: Measure of efficiency

Collaboration success: It is one of the political goals of the ESPRIT program to bundle the competencies from different countries and to support technological exchange between universities and industry. Our indicators assigned to the construct collaboration success assess to what extent the participants have become acquainted with new, valued partners, to what extent they plan to continue collaboration with these new partners beyond the end of the project, and to what extent they plan to carry out a joint project with these new partners even without public funding.

Indicator	Item-to-total correlations	Factor loadings
During the project, your company has become acquainted with new, valued partners.	.60	.82
Your company has plans to continue collaboration with these new partners beyond the end of the project.	.74	.89
Your company has plans to carry out a joint project with these new partners even without public funding.	.64	.84
Summarizing statistics Cronbach's alpha / explained variance	.81	72.3 %

Table 4: Measure of collaboration success

Learning success: An often-mentioned goal in the cooperation literature is to learn from partners how they develop and implement new products or processes. This means technical know-how and engineering practices, social skills, and practices which have a long tradition in other countries and cultures. Our indicators measure these different aspects of learning. The following documents the results of our scale analysis. Although Cronbach's alpha falls slightly below the .70 limit, explained variance, Item-to-Total correlations and factor loadings show sufficiently high values. We can use the two items to measure learning.

Indicator	Item-to-total correlations	Factor loadings
Through the collaboration, your company has learned technical modes of operation and methods of the partners.	.48	.86
Through the collaboration, your company has learned to efficiently work with European partners.	.48	.86
Summarizing statistics Cronbach's alpha / explained variance	.65	74.2 %

Table 5: Measure of learning success

3.2.2 Measure of the starting conditions

Competence: For the measurement of the multi-faceted construct *competence*, which plays a central role in the resource-based view of the firm, we used an extensive battery of items. Our statements include overall assessments of technological competence and the capabilities of all partners to fulfill the requirements of the project, and specific questions regarding the competencies of the developers and users which participated in the project. Three other items measure the perceived qualities of references, infrastructure and employees.

Indicator	Item-to-total correlations	Factor loadings
All partners were able to fulfill the requirements of the project.	.47	.51
The users in the project offered technologically leading products or services.	.67	.70
The users in the project pursued the strategy of a leading technology user.	.64	.70
The users in the project could specify their required functionality sufficiently.	.58	.63
The developers were technologically leading in comparison with their European competitors.	.67	.73
The developers in the project pursued the strategy of a technology leader.	.45	.57
The partners had qualified employees.	.69	.79
The partners had an adequate infrastructure.	.44	.57
The partners had excellent references.	.61	.74
Overall, at project start, the consortium was represented by technologically competent partners.	.69	.79
Summarizing statistics Cronbach's alpha / explained variance	.86	46.0 %

Table 6: Measure of competence

Cronbach's alpha and the item-to-total correlations show high values. The explained variance lies below 50%, but it is very difficult to explain the common variance of ten items with only one factor. Given the large number of items, there still remains substantial common variance. The factor loadings of the items also document a solution with only one factor. We therefore use all ten items to measure the competence of the R&D consortium.

Complementarity: While competence is necessary to complete a task, *complementarity* of the resources of different partners is the critical motivator for them to cooperate. Our statements include different aspects of resource complementarity. We measure whether the contributions of all partners were necessary for the project.

Indicator	Item-to-total correlations	Factor loadings
The contributions of all partners were necessary for the project.	.57	.71
The competencies of the partners compelemented each other adequately.	.71	.83
Due to the complementary contributions of the partners, considerable synergies were expected.	.74	.85
The pursued project results required an intensive collaboration among partners.	.64	.77
Overall, at project start, the input resources of the consortium complemented each other well.	.75	.86
Summarizing statistics Cronbach's alpha / explained variance	.86	65.2 %

Table 7: Measure of complementarity

Goal clarity: Our items for measuring *goal clarity* include statements about desired functionality of the prototype and its performance characteristics, clarity and measurability of the project objectives and the contributions of the individual partners.

Cronbach's alpha, explained variance, item-to-total correlations, and factor loadings show very high values. We can use the five items to measure the goal clarity of the cooperative project. It is worth regarding the legal issues of exploitation of the gained know-how does not load sufficiently on our factor goal clarity. This confirms our choice to restrict the measurement of effectiveness on the achievement of technical objectives: if commercial objectives substantially lack clarity, then we have no standard against which to compare the actual results.

Indicator	Item-to-total correlations	Factor loadings
The partners knew exactly, which objectives were to be reached in the project.	.76	.85
The desired functionality of the prototype was set in detail.	.84	.90
Performance characteristics of the prototype were defined by testable limits.	.81	.88
Required performance contributions of individual partners were defined precisely.	.75	.84
Overall, at project start, the objectives were defined clearly and verifiable.	.85	.91
Summarizing statistics Cronbach's alpha / explained variance	.92	76.5 %

Table 8: Measure of goal clarity

Goal compatibility: Cooperative R&D projects do not only promise rewards. They also have many inherent risks, and it takes additional effort to establish a well-functioning cooperation. Usually, the partners of a cooperation do not always share common interests, and conflicting goals are possible as well. However, if the conflicts are too strong, they may escalate during the project and cause detrimental effects. We measured the extent of goal compatibility at the start of the project by asking the respondents about technical objectives, distribution of budgets and work packages, and exploitation of results. We also asked to what extent some partners used pressure to push their objectives, and to what extent some partners imposed their objectives at the expense of others. All items are reverse coded because we want to measure goal compatibility and not goal conflict.

Cronbach's alpha and the item-to-total correlations show high values. The explained variance just falls short of the 50 % limit, but it is very difficult to explain the common variance of eight items with only one factor. Given the large number of items, there still remains substantial common variance. The factor loadings of the items also document a solution with only one factor. Hence we use all ten items to measure the goal compatibility of the R&D consortium.

Indicator	Item-to-total correlations	Factor loadings
There were different points of view on the major technical objectives.	.59	.71
There were different points of view on the extent of the pursued application orientation.	.59	.69
There were different points of view on the distribution of the budget.	.55	.64
There were different points of view on the distribution of work packages.	.57	.61
There were different points of view on the exploitation of results.	.62	.70
Some partners imposed their objectives at the expense of others.	.68	.78
Some partners used pressure to push through their objectives.	.62	.72
Overall, at project start, there were different points of view on the project objectives.	.67	.76
Summarizing statistics Cronbach's alpha / explained variance	.86	49,4 %

Table 9: Measure of goal compatibility

Trust: In a recently published comprehensive review on the research about trust, Rousseau, Sitkin, Burt, and Camerer (1998, p. 395) write: "*Trust is a psychological state comprising the intention to accept vulnerability based upon positive expectations of the intentions or behavior of another.*"[4] In our approach to measure the level of trust which exists in the R&D consortium, we ask the participants of the projects to what extent the bases for positive expectations about the partner's intentions or behaviors are given. This means that we ask the respondents to what extent they perceive their partners as reliable, benevolent, honest and open, and we ask for

[4] For similar definitions see also: Butler 1991; Morgan, Hunt 1994; Mayer, Davis, Schoorman 1995; Plötner 1995; Fennetau, Guibert 1997; Helfert 1998, p. 116 ff.; Smith, Barclay 1997; Saad 1998, p. 102 ff.; Walter 1998, p. 222 ff.

typical behaviors expressing such characteristics. We assess whether there was a communicative atmosphere in the consortium and whether the partners were willing to share information relevant for the project, whether important information was concealed, whether one could depend on the correctness of the supplied information, whether reached decisions were not challenged afterwards, whether given commitments were kept, and whether one could depend on agreements of confidentiality. The item battery also contains an overall question about the basis of trust among the partners, in order to test whether the assumed conditions of trust are indeed positively linked to trust as a psychological state. Overall content validity of our scale should be high.

Indicator	Item-to-total correlations	Factor loadings
At meetings held before project start, there was a communicative atmosphere.	.57	.71
The partners were willing to supply information relevant for the project.	.66	.78
Important information for the project was not concealed.	.67	.74
One could depend on the correctness of the supplied information.	.74	.84
Reached decisions were not challenged afterwards.	.71	.77
Given commitments were kept.	.68	.75
One could depend on sensitive information not being misused.	.61	.70
Overall, at project start, there was a basis of trust among the partners.	.72	.81
Summarizing statistics Cronbach's alpha / explained variance	.88	58.2 %

Table 10: Measure of trust

Commitment: Moorman, Zaltman, and Desphandé (1992, p. 316) define commitment as: *"an enduring desire to maintain a valued relationship"*.[5] Applied to R&D cooperations, this means that the partners are highly motivated to make the project a success and are therefore willing to invest time and money in adequate preparation prior to the project's start. Hence we measure commitment by asking questions regarding the partner's willingness to invest time, money, and extra effort into a good proposal. The item battery also contains an overall statement about the willingness of the partners to meet their obligations.

Indicator	Item-to-total correlations	Factor loadings
The partners invested much time and money to build up the consortium.	.54	.77
The partners invested much time and money to develop a good proposal.	.64	.84
The partners were willing to contribute to necessary additional tasks.	.47	.63
Overall, at project start, there was a high willingness to meet the incurred obligations.	.43	.69
Summarizing statistics Cronbach's alpha / explained variance	.73	54.1 %

Table 11: Measure of commitment

3.2.3 Measure of project progress

Escalation of conflicts: This construct is measured by the two items "It frequently came to conflicts" and "The degree of conflict escalation was high". They correlate at .87 and can thus be combined to a scale with a Cronbach's alpha of .93. Both factor loadings are at .97, the explained variance is 93.3 %.

---

[5] For a discussion of the construct and its measurement see also: Söllner 1993; Morgan, Hunt 1994, p. 23; Helfert 1998, p. 116 ff.; Saad 1998, p. 109 ff.; Walter 1998, p. 224 ff.).

Changes of goals: This construct is measured by the two items "The project objectives were often changed" and "At least one major project objective was changed substantially". They correlate at .61 and can thus be combined to a scale with a Cronbach's Alpha of .75. Both factor loadings are at .90, and the explained variance is 80.7 %.

Project management: Our measurement of quality of *project management* comprises two aspects which have been shown in other investigations to be critical success factors of various kinds of projects: The quality of communication and the quality of planning and controlling.[6] The three items measuring communication quality ask for the know-how to handle crises during the project, for the pre-definition of communication channels, and for the information about the actual state of the work. The items that are used to measure planning and controlling assess the extent to which budgets, schedules, resources, and performance progress are monitored, the revision and adaptation of plans, if necessary, and the quality of the project management procedures and tools.

Indicator	Item-to-total correlations	Factor loadings
Everybody in the consortium knew exactly where help was available when difficulties arose.	.63	.73
The communication channels in the consortium were predefined.	.68	.77
All partners were always sufficiently informed concerning their work on the project.	.76	.84
All essential aspects of the project execution were monitored and controlled (budget, time schedule, resources, performance progress).	.80	.88
During the project, planning data were reviewed and adapted if necessary.	.73	.83
Methods, procedures and tools for project control and supervision worked well.	.83	.90
Summarizing statistics Cronbach's alpha / explained variance	.90	68.1 %

Table 12: Measure of project management

---

[6] cf. Lechler 1997; Gemünden, Lechler 1997; and Lechler, Gemünden 1998 for a review of existing empirical studies and a new core model of the success factors of project management.

3.2.4 Summary of the scale analyses

The following table summarizes the results of our scale analyses. Overall, the scales show good values for convergent validity and can be used for further analysis.

Construct	No. of items	Cronbach's alpha	Variance explained
Overall success	4	.90	76.0 %
Effectiveness (output quality)	5	.94	79.3 %
Efficiency (budget & schedule)	2	.77	81.9 %
Collaboration	3	.81	72.3 %
Learning	2	.65	74.2 %
Competence	10	.86	46.0 %
Complementarity	5	.86	65.2 %
Goal clarity	5	.92	76.5 %
Goal compatibility	8	.86	49.4 %
Trust	8	.88	58.2 %
Commitment	4	.73	54.1 %
Escalation of conflicts	2	.93	93.3 %
Changes of goals	2	.75	80.7 %
Project management	6	.90	68.1 %

Table 13: Summary of measures

4. Empirical results

In order to test our hypotheses we first analyze the bivariate correlations of the starting conditions and the project progress variables with the five dimensions of project success. We then perform path analyses to investigate the multivariate influence. In doing so, we first estimate two basic models. The first describes the influence of the core starting conditions (competence, goal clarity, and trust) on project success. The second describes the influence of the project progress variables on project success. Lastly, we will estimate an extended model performing a path analysis with all core constructs.

4.1 Bivariate analysis

The following table shows the bivariate correlations of the starting conditions and the project progress variables with project success. All independent variables show a substantial and highly significant correlation with the three typical success dimensions of a project, i.e., overall success, effectiveness, and efficiency ($p \leq 0.001$). This confirms our respective hypotheses and documents that each of the starting conditions delivers valuable information about project success. We can also see that project management is an important success factor and that escalation of conflicts and changes of goals should be avoided. (For similar findings see Lechler 1997, p. 258 ff., and Lechler, Gemünden 1998).

The correlations shown above are not equally high. Goal compatibility shows a much lower correlation than the other variables. It appears to be a less important starting condition. It is very likely that if the participants perceive too high goal conflicts they will not engage in a cooperative project, because they perceive that the risk of failure is too high. Therefore we only observe projects with low and moderate initial goal conflicts.

	Overall success	Effectiveness (quality of output)	Efficiency (time and budget)	Collaboration success	Learning success
Competence	.57	.49	.44	.41	.38
Complementarity	.42	.34	.37	.23	.30
Goal clarity	.43	.48	.36	.28	.20
Goal compatibility	.26	.26	.26	.17	.08
Trust	.51	.42	.45	.33	.36
Commitment	.43	.37	.36	.28	.39
Escalation of conflicts	-.45	-.45	-.41	-.33	-.17
Goal changes	-.39	-.39	-.41	-.30	-.20
Project management	.44	.40	.58	.32	.34

Table 14: Bivariate analysis

Competence, trust, and goal clarity, which we assumed to be the most important variables, show the highest correlations with the success variables, although the correlation coefficients of complementarity and commitment are not much lower.

The correlations with collaboration and learning are in most cases lower than for overall success, effectiveness, and efficiency.

Looking at *learning*, it appears that the participants often nearly learned as much in cases of unclear, changing, and conflicting goals and escalating conflicts, as in cases of clear, stable, and compatible goals and the absence of conflicts. There could be different reasons for this. Projects with unclear and conflicting goals may be of a particular kind: They involve more participants, have a longer duration, are dealing with more basic research questions and more uncertain problems. Hence it could have been more difficult to establish clear goals (Hauschildt 1977, p. 47 ff.) - it is one of the crucial tasks of the project to find the critical issues instead of postulating a clear but wrong mission. The second reason may be that the participants were ill prepared. They did not know each other as well as the members of other consortia, they did not invest enough time and work to find adequate partners and formulate a better proposal. The third reason may be that the participants recognize that their project has not been a success in terms of effectiveness, efficiency, and overall objectives. Consequently, they emphasize learning effects, so that this project had at least one 'good outcome' to offset the resources invested. However, it is not at all a given fact that we only learn from mistakes and failures. Project consortia with a higher degree of competence and complementarity, and with a higher level of trust and commitment, also show higher levels of learning. It appears that these project members have profited from more open communication with higher-valued and more knowledgeable partners so that they acquired more know-how. Moreover, these partners were more satisfied with their results and the social recognition they experienced. This is very different from learning from failures, where the project members experience distress and frustration.

Collaboration is more positively influenced by clear and compatible goals than learning. It is also more negatively influenced by escalating conflicts and changing goals, but the influences are still lower than for the typical success dimensions (i.e., overall success, effectiveness, and efficiency). It might be that the acquisition of new partners is not the prime goal of the cooperative project. The main task is to fulfill the requirements of the contracted goals within the given time and budget constraints. Although, some participants wanted to become acquainted with new cooperation partners, they still must contribute their share to the project.

4.2 Path analysis of the influence of the core starting conditions on project success

For our path analysis of the starting conditions and project success, we have concentrated on the three theoretically and empirically most important constructs: competence, goal clarity, and trust. Since we have only a sample of 167 cases, we use fewer indicators to measure the constructs. We use the sum of the scores of overall success, effectiveness, and efficiency as indicators of the latent variable project success. The latent variables competence, goal clarity, and trust are also measured with only three to five indicators in order to save parameters and to estimate a more robust model. This reduction of the indicators also secures a higher discriminant validity of our measurement. All path models are estimated with the AMOS software using the ULS estimator. All paths shown in the models below are significant at least at the 5 % level.

Figure 2 shows the result of our analysis of the core starting conditions and project success. Competence shows substantial positive influences on goal clarity and trust, while holding goal clarity con-

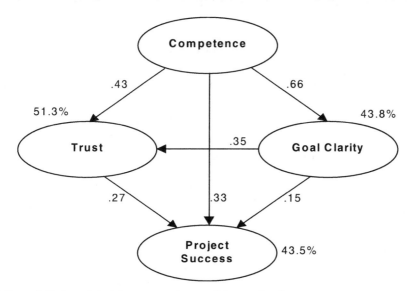

Figure 2: Path model of the core starting conditions and project success

stant. Each of the three starting conditions also has a positive influence on projects' success, with competence having the strongest influence followed by trust and goal clarity. Thus, all hypotheses of our basic model are confirmed. Together the three starting conditions explain 44 % of the variance of project success.

Table 15 shows the direct, indirect, and total effects of the three starting conditions on project success. With a strong total effect of .60, *competence is the most important starting condition*. With total effects of .24 and .27, goal clarity and trust have about the same medium-size influence.

	Direct effect	Indirect effect	Total effect
Competence	.33	.27	.60
Goal clarity	.15	.09	.24
Trust	.27	-	.27

Table 15: Total effects of the path model of the core starting conditions and project
      success

4.3 Path analysis of the influence of the project progress on project success

Figure 3 shows the result of our path analysis for the project progress variables and project success. Project management quality shows a substantial negative influence on escalation of conflicts and a much smaller negative influence on changes of goals. Since these two syndrome variables both have negative influences on project success, higher project management quality contributes to a reduction of negative consequences, thus positively impacting project success. Project management quality also has a direct positive influence on project success. Finally, conflict escalation has a negative influence on goal changes, which means that escalating conflicts also have a negative indirect influence.

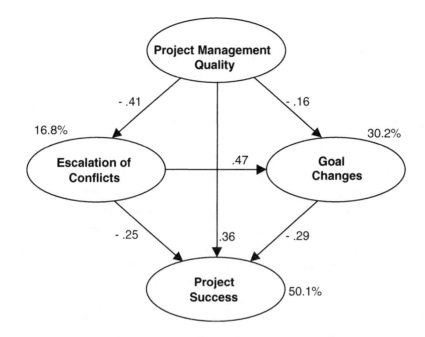

Figure 3: Path analysis of the influence of the project progress on project success

Table 16 summarizes the direct and indirect influences. Project management quality has a strong positive influence on project success (+.57). With total effects of -.39 and -.27, escalation of conflicts and changes of goals exert substantial negative influences on project success. All hypotheses of our basic model are confirmed. Together, the three progress variables explain 50% of the variance of project success. This is a considerable amount.

	Direct effect	Indirect effect	Total effect
Project management quality	.36	.21	.57
Escalation of conflicts	-.25	-.14	-.39
Changes of goals	-.29	-	-.27

Table 16: Total effects in the path analysis of the influence of the project progress on project success

4.4 Path analysis of the full model

Figure 4 shows the result of our path analysis on simultaneous influences between the core starting conditions, project progress variables, and project success (full model). Project management quality still has a direct positive influence on project success. However, holding trust and goal clarity constant, it has no more direct influence on conflict escalation and goal changes. Conflict escalation is strongly reduced by trust, and changes of goals are somewhat lower if initial goal clarity is higher. Conflict escalation still strongly increases goal changes, but the direct influence of conflict escalation on project success is no more significant. The interrelationships between all three starting conditions are still significant and have about the same size as in the basic model of (only) the starting conditions. However, whereas competence still has a substantial direct effect on project success, trust and goal clarity now show only indirect effects. Project management quality is strongly increased by trust and moderately increased by goal clarity, but competence has no direct influence (only indirect influences). Project management quality is based on communication quality. Consequently trust between the partners appears to be a crucial condition for project management quality. Together, the six variables explain 65% of the variance of project success.

Table 17 summarizes the direct and indirect influences. Competence still shows by far the largest total effect. The total effects of trust, project management quality, and changes of goals are substantial, but much lower. Goal clarity and escalation of conflicts show a small influence.

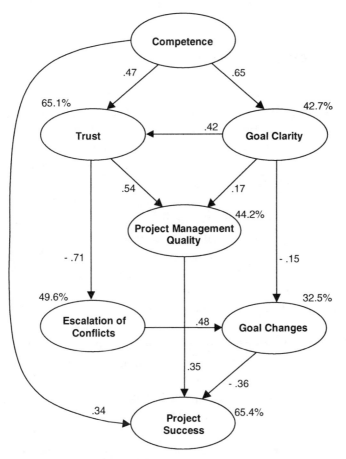

Figure 4: Path analysis of the full model

	Direct effect	Indirect effect	Total effect
Competence	.34	.30	.64
Goal clarity	n.s.	.24	.24
Trust	n.s.	.31	.31
Project management quality	.35	n.s.	.35
Escalation of conflicts	n.s.	-.17	-.17
Changes of goals	-.36	-	-.36

Table 17: Total effects in the full model.

5. Discussion and outlook

We have developed a conceptual model of starting conditions comprising resource fit, social fit, and goal fit, which should be fulfilled in international R&D cooperations. Additionally, we have proposed reliable and valid measures for our constructs. All starting conditions show substantial positive correlations particularly with the three core project success variables - overall success, effectiveness, and efficiency. Among the six starting conditions - competence, complementarity, goal clarity, goal compatibility, trust, and commitment - competence and trust appear to be the two most important. Additional regression analyses document that these two variables account for most of the variance of project success, which can be explained by the starting conditions individually.

By introducing three variables of project progress - quality of project management, escalation of conflicts, and changes of goals - we were able to explain 65% of the variance of project success and show how the starting conditions exert their influence.

This means that our findings are useful for the participants who want to form a consortium, for the political decision makers and reviewers, and for the project consultants involved. The European Union can increase the effectiveness of its program by using the starting conditions as selection criteria. The participants should consider that it is not sufficient to write a good proposal that secures co-funding from the European Union. ESPRIT projects go through a long and highly selective initiating phase with sometimes unacquainted partners and relatively few meetings. But still, there are quite significant investments to be made by the partners in order to craft a detailed and sound project proposal and to negotiate with the European Commission, before an ESPRIT project can officially be launched. The high rate of rejected projects and the budget cuts document that the investments are not without risk. However, our study also demonstrates that even if these risks have been dealt with, the funded projects may fail because of inadequate starting conditions. This highlights the necessity for the partners to pay special attention to the starting conditions.

We have to mention that our investigation is not free of methodological issues. The starting conditions, the progress variables, and the project's success were evaluated after the completion of the projects. Thus, a positive or negative experience of a project's progress may have distorted the ex post perception of the starting conditions. It is possible that the explained variance of project success is somewhat overestimated. However, in our view the amount of explained variance is so high that a substantial amount will still remain after subtracting method influences. The results are very promising and justify a longitudinal study, which would, however, require even more resources than this large-scale empirical investigation.

Future research should not only perform a longitudinal test of our model. It could also address interesting other questions:

1. How do the participants establish a consortium with good starting conditions? What does the ideal initiating phase look like? How does one seek and select appropriate partners?

2. Are the starting conditions equally important for all kinds of projects? We expect that for projects with a shorter duration the influence of starting conditions will be stronger, since it is more difficult to compensate inadequate starting conditions in a shorter period of time. In addition, many unexpected events can happen in projects that span several years. We expect for consortia with many partners that starting conditions will have a weaker impact on the ultimate success of the project. This stems from our belief that it will be more difficult to sustain a high level of trust, commitment, goal clarity, and goal consensus in a larger group of partners.

3. Are the starting conditions conceptualized here also important for the progress and success of the exploitation activities? How do the starting conditions influence commercial success that results from cooperative European R&D projects?

Acknowledgements

The authors are grateful for the financial support from the State of Baden-Württemberg, Federal Republic of Germany. They would also like to thank Peter Lockemann, Monika Bachofner and Bernhard Kölmel for their work in this research project and Wesley Urquhart for his help in translating and editing this paper. The first author dedicates this article to his friend Monika Hachmann.

Concurrent development and product innovations

Eric H. Kessler / Alok K. Chakrabarti

Concurrent development and product innovations

Eric H. Kessler / Alok K. Chakrabarti

1. Introduction

Concurrent development, or simultaneously executing multiple stages of the product development process, is regarded by many as an efficient and effective method for increasing the speed of innovation (see Figure 1) while also providing a means for containing development costs and improving the quality of the end product (Gilbert 1995; Zairi, Youssef 1995). It has also been referred to as parallel versus linear processing of tasks (Millson et al. 1992), co-ordinating efforts rather than "throwing the product over the wall" from one stage to another (Brown, Karagozoglu 1993), and as a "rugby" method of constant, multi-disciplinary team interplay rather than a "relay race" method of phase-to-phase progression with func-tionally specialized and segmented divisions (Smith, Reinertsen 1991; Souder, Chakrabarti 1980; Takeuchi, Nonaka 1986). Consis-tently, the Institute for Defense Analysis (Handfield 1994) refers to concurrent development "as a systematic approach to the integrated concurrent design of products and related processes including manu-facture and support ... [which] causes the developers, from the out-set, to consider all the elements of product life-cycle from concep-tion through disposal including quality, cost, schedule, and user requirements."

However, alternative arguments exist in the literature which posit negative effects of concurrent development on these performance dimensions, for instance due to reduced control over the process. For example, concurrent development might drive up costs due to the need to create and maintain more complex communication networks (Graves 1989). Additionally, concurrent development might lower

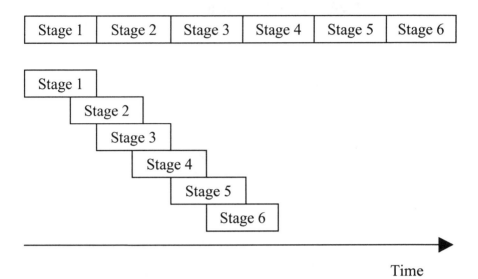

Time

Figure 1: Contrast between non-overlapping (sequential) and overlapping (parallel) product
development processes plotted against time of development

product quality because the extra pressure put on workers may cause
more mistakes and allow for insufficient time to adapt to the market
(Crawford 1992; Handfield 1994). To date, it is unclear which of
these perspectives is correct, for the empirical evidence regarding
outcomes of concurrent development is inconclusive (Handfield
1994). Thus the research questions we address are threefold:

(1) How does concurrent development influence process speed?

(2) How does concurrent development influence development
    costs?

(3) How does concurrent development influence product quality?

2. Hypotheses

The conceptual mode illustrated in Figure 2 is used as a basis for
empirically testing the three research questions. Given the exploratory nature of the research, hypotheses will adopt the falsifiable
assumption that concurrent development has generally positive
effects on speed, quality, and costs. That is, they assume that there
are no trade-offs between the three dimensions of product development.

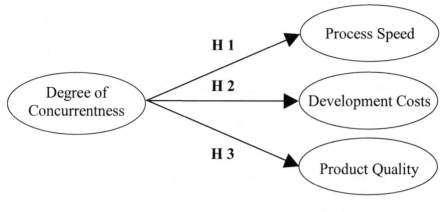

Figure 2: Conceptual model of performance outcomes of concurrent development

2.1 Concurrent development and speed

Concurrent engineering, by allowing the different tasks of the product development process to be carried out simultaneously rather than sequentially, allows developers to consider all the elements of product life-cycle from conception through market launch (Gilbert 1995; Izuchukwe 1992; Kessler, Chakrabarti 1996; Zairi, Youssef 1995). It is thus considered one of the most effective facilitators of innovation speed. This is evidenced by empirical research in the new product development literature (e.g., Terwiesch, Loch 1998) as well as observations in the logistics and operations management literature describing developers' use of computer programs to identify the contribution of overlapping activities to reducing critical paths and slack times (Zhu, Heady 1994). Conversely, a lack of overlap can waste time by forcing downstream tasks to wait for previous stages to be completed in their entirety, thereby lengthening the critical path of projects. It also limits the communications between functions, increasing the number of time-consuming design changes in the production phase of product introductions (Deschamps, Nayak 1992; Vesey 1991). This is because information is communicated in periodic "batches" (versus continuously), necessitating longer time periods to assimilate the information (Blackburn 1992; Clark, Fujimoto 1991; Rosenthal 1992). Thus we propose the following:

HYPOTHESIS 1:
A greater degree of concurrent development will be related to faster new product development.

2.2 Concurrent development and costs

Concurrent engineering seeks to optimize the design and manufacturing process to reduce cost by the integration of activities and parallelism in work practices (Zairi, Youssef 1995). In this sense, it can be viewed as a means toward design efficiency (Reed et al. 1996). Parallel processes can reduce overall development time, thereby helping to cap man-hours (Rosenthal 1992), increase the efficiency of resource utilization (Clark 1989), and reduce costly work redundancy, errors, and recycling (Meyer 1993; Rosenau 1988). Indeed, evidence from the Ford Taurus development project supports the cost-saving potential of concurrent engineering (Gilbert 1995). Moreover, combining concurrent engineering processes with in-process design controls and computer and information technology was recently shown to significantly reduced the costs of product development (Hull et al. 1996). Thus we propose the following:

HYPOTHESIS 2:
A greater degree of concurrent development will be related to lower new product development costs.

2.3 Concurrent development and quality

Concurrent development can also help in increasing the quality of the end product developed, defined as the degree to which it satisfies customer requirement (Clark, Fujimoto 1991). This is because the integration of activities and parallelism in work practices which come from concurrent development tend to facilitate higher rates of communication, learning, and problem solving among project team members (Ancona, Caldwell 1990; Eisenhardt, Tabrizi 1995; Meyer 1993). A more compact process may also improve forecasting and market-fit, primarily because firms are required to project competitor movements, developments in technologies, and demographic trends into shorter time periods (Deschamps, Nayak 1992; Page 1993; Wheelwright, Clark 1992). Indeed, concurrent engineering is considered a key component of an integrated total quality management (TQM) program (Reed et al. 1996). Thus we propose the following:

HYPOTHESIS 3:
A greater degree of concurrent development will be related to higher quality of the new product developed.

3. Methodology

3.1 Sample

Selection of the research sample is motivated by the objective of being able to generalize the findings of this study beyond (a) the idiosyncratic nature of undeveloped, unconventional product development programs and instead across organizational boundaries, and, (b) the idiosyncratic nature of one or two task/institutional environments and instead across industry boundaries. As a result, the sample consists of large (greater than $50 million in sales) U.S.-based companies in several industries. Large firms were chosen because they are more likely to have established new product development programs as opposed to smaller firms with more idiosyncratic programs. Firms in different industries were chosen because they provided access to a range of environments where product innovation is pursued and hence allowed the study to more broadly examine the implications of concurrent development.

Given these criteria, company names were assembled in a systematic manner following the site selection algorithm developed by Souder and Chakrabarti (1980). Thirty (30) companies were chosen which met the criteria of the study and were headquartered locally, which was a practical research constraint (e.g., travel resources). Site entry letters were sent to the chief executive officer (CEO) or top research and development executive (e.g., VP, R&D) of these firms to provide a general overview of the study, explain the nature of the commitment requested, describe the study's confidentiality policy, and detail the benefits of participation. Two to three weeks later, direct telephone calls were made to these individuals to answer any questions they had about the study and arrange a mutually convenient time for an onsite interview (to secure commitment).

Ultimately, this procedure resulted in ten companies agreeing to participate in the study. The participating companies operated in a variety of industries and had an average of 89,662 employees and an average of $16,014.36 million in sales. Because a variety of industries were sampled, findings are more generalizable than single-industry studies and indicate more general principles of concurrent development which apply across several product types and industrial environments.

The unit of analysis was the new product development project, defined as "a goal directed effort with a readily-identified end in view" (Rubenstein et al. 1976). This enabled us to capture the multiple attributes of actual projects. A total of 86 projects were selected from the ten firms; of this population, questionnaires representing 75 projects were returned (87% response rate). Multiple respondents were polled from each project to increase the validity and reliability of their reports (Kumar et al. 1993). A total of 205 individuals from the 86 projects were identified as potential respondents; of this population, 127 surveys were returned (62% response rate).

3.2 Measures

The four primary variables in the study are concurrent development (i.e., project overlap), innovation speed, development costs, and product quality. They were operationalized in the following manner.

Concurrent Development. The degree to which stages of the development processes were undertaken in parallel was calculated as the sum of the time in months of the stages of the product development project divided by the total product development time. The stages, adopted from Eisenhardt and Tabrizi (1995) are:

(a) Pre-development/planning, which begins with the start of the project and ends with the completion of basic product requirements;

(b) Conceptual design, which begins with the basic concepts and ends with final specifications of the product;

(c) Product design, which begins with the engineering work to take the specifications to a fully designed product and ends with final release to system test;

(d) Testing: begins with component and system test and ends with the release of the product to production;

(e) Process development, which begins with the first process design and ends at the completion of the first pilot run; and

(f) Production start-up: begins with production ramp-up and ends with the stabilization of production.

For example, if the project was undertaken sequentially, the sum of the stage times would equal the total time. Alternatively, if the project was undertaken in parallel (i.e., two or more stages overlapped), the sum of the stage times would be greater than the total time. A higher score indicates a higher degree of project overlap.

Innovation Speed. Speed was operationalized through three relative measures:

(a) speed relative to schedule, or on-time performance (McDonough 1993; Mc Donough, Barczak 1991),

(b) speed relative to similar, previously completed projects in one's organization, or acceleration (Crawford 1992; Millson et al. 1992; Nijssen et al. 1995), and

(c) speed relative to similar projects of competitors, or competitive speed (Birnbaum-More 1990; Vesey 1991).

For each scale, respondents were asked to check off one of 13 boxes describing projects as relatively faster, slower, or equal in speed to schedules, past projects, or competitor projects. The three scales had a moderate degree of internal reliability (a = .6824). When the scales were subjected to principle component factor analysis, they all loaded onto one factor with eigenvalues equal to or greater than one. Thus a weighted aggregate innovation speed variable was derived ranging from one (lowest time / fastest speed) to thirteen (highest time / slowest speed).

Development Costs. Development cost was also measured in three ways, which mirrored the measurement of innovation speed. A project's cost relative to its budget was measured relative to

(a) budget,

(b) similar past projects, and

(c) similar competitor projects.

For each scale, respondents were asked to check off one of 13 boxes describing projects as relatively less expensive, more expensive, or equal in cost to schedules, past projects, or competitor projects. The three scales had a moderate degree of internal reliability (a = .6173). When the scales were subjected to principle component factor analysis, they all loaded onto one factor with eigenvalues equal to or greater than one. Thus a weighted aggregate development cost variable was derived ranging from one (lowest cost) to thirteen (highest cost).

Product Quality. Product quality was also measured in three different ways, again mirroring the measurement of innovation speed. Quality was measured relative to

(a) preset performance specifications,

(b) similar past projects, and

(c) similar competitor projects.

For each scale, respondents were asked to check off one of 13 boxes describing projects as relatively higher quality, lower quality, or equal in quality to schedules, past projects, or competitor projects. The three scales had a moderate-to-high degree of internal reliability (a = .7825). When the scales were subjected to principal component factor analysis, they all loaded onto one factor with eigenvalues equal to or greater than one. Thus a weighted aggregate product quality variable was derived ranging from one (lowest quality) to thirteen (highest quality).

3.3 Analysis

We sought to answer the research questions in two ways. First, and most simply, by testing how important concurrent development is when used alone to predict cost, quality, and speed. The answer to this question is gained by looking at correlation coefficients (SPSS 1994). Thus, bi-variate correlation coefficients were calculated for concurrent development and each of the dependent variables - cost, quality, and speed.

Second, and perhaps more realistically, we tested how well concurrent development predicts cost, quality, and speed when considered along with other relevant independent variables in the innovation process[1]. The answer to this question is gained by looking at regression equations (SPSS 1994). This answer is more complex, however, because any statement about concurrent development is contingent upon the other variables considered in the model. Thus,

---

[1] We used the model of innovation antecedents developed by Kessler and Chakrabarti (1996) to motivate variable selection. See Table 3 for a complete list of independent variables in the empirical model.

for each of the dependent variables, an automatic search procedure (or stepping procedure) was used to develop a best (or parsimonious) subset of independent variables to be included in the regression model (Neter et al. 1990). There are several, largely similar analytical approaches to the task of sequentially analyzing all antecedent factors while adjusting for the affects of one another on innovation speed. SPSS (1994) reports that none of these selection procedures are best in any absolute sense. Instead, the choice between approaches should be made on the basis of the objectives of the analysis.

Because we desired to test the influence of the independent variables while simultaneously controlling for the others, backward-elimination was chosen - backward-elimination regression analysis allows the researcher to examine each independent variable in the regression function adjusted for all the other independent variables in the pool (Neter et al. 1990). Specifically, backward-elimination of independent variables starts with all the variables in the equation (as opposed to forward-selection and stepwise-selection, which add variables one at a time) and sequentially removes them. The removal criteria used are probability of F-to-remove (POUT). The POUT procedure removes variables from the regression equation sequentially, beginning with the variable with the highest p-value, and continues to recompute the regression equation and remove variables until all remaining variables have a p-value of less than 0.10 (SPSS default criterion). The final equation represents the parsimonious (or best) regression model.

4. Results

Table 1 reports the descriptive statistics for concurrent development, speed, costs, and quality. Results revealed that all variables had generally broad ranges and central means. Further, tests revealed low degrees of kurtosis (i.e., spiking) and skewness (i.e., imbalance) in their distributions which suggest that there is no significant threat to the assumption of normality.

Table 2 reports the results from bi-variate correlations. Results revealed that, by itself, concurrent development is a poor predictor of cost and quality and only a marginal predictor of speed ($p < .10$).

Variable	Scale	Mean	SD	Max	Min
Concurrent Development	Ratio	2.08	0.79	4.00	1.00
Speed	1-13	5.35	1.39	9.45	2.59
Costs	1-13	6.06	1.21	8.70	2.64
Quality	1-13	7.30	1.38	10.65	4.34

Table 1: Descriptive statistics for concurrent development, costs, quality, and speed

Dependent Variable	r	Sig r
1. Speed	.2437	$p < .10$
2. Costs	.1102	ns.
3. Quality	-.1222	ns.

Table 2: Bi-variate correlations of concurrent development with costs, quality and speed

Table 3 reports the results for the three backward-elimination regression procedures examining the effects of several potential antecedents on each of these dimensions. Results revealed that

(a) concurrent development was a significant factor in all three parsimonious models, and

(b) concurrent development had *mixed* effects on product innovation outcomes - a higher degree of concurrentness was found to increase process speed ($p < .01$), but it also tended to increase development costs ($p < .01$) and decrease product quality ($p < .05$).

Dependent Variable	B	SE B	Beta (β)	T	Sig T
1. Speed Full Model F = 6.00 (p<.001) Full Model R^2 = 0.44	1.066	0.295	0.601	3.581	>.001
2. Cost Full Model F = 3.90 (p<.001) Full Model R^2 = 0.61	0.998	0.284	0.650	3.507	.001
3. Quality Full Model F = 3.81 (p<.01) Full Model R^2 = 0.38	-0.489	0.230	-0.281	-2.130	.039

Table 3: Effect of concurrent development on costs, quality, and speed taken from three separate backward-elimination regression analyses

Table 4 reports the correlations between concurrent development and the other independent variables inputted into the backward-elimination regression procedures. Results revealed that concurrent development was associated with

(1) greater emphasis placed on speed (versus costs or quality) by top management (p<.05),

(2) greater reward system orientation for speed (p<.10),

(3) lower clarity (i.e., more ambiguity) surrounding the project goals (p<.01),

(4) lower clarity (i.e., more ambiguity) surrounding the product concept (p<.05),

(5) more radical innovations (p<.10),

(6) lower project member tenure (p<.01),

(7) lower authority (i.e., less empowerment) of project teams (p<.05),

(8) greater turfguarding (p<.05),

(9) greater frequency of testing (p<.01), and

(10) lower use of CAD systems (p<.01).

Strategic Orientation - Criteria	
1 Importance of Speed	.2714 *
2 Reward for Speed	.2263 ᵞ
3 Culture Orientation	-.0568
4 Time-Goal Clarity	-.3734 **
5 Product-Concept Clarity	-.2969 *
6 Top Management Interest in Project	-.2085
Strategic Orientation - Scope	
7 Project Stream Breadth	.1134
8 Product Radicalness	.2343 ᵞ
9 External Sourcing of Ideas / Technologies	-.1694
Organizational Capability - Staffing	
10 Product Champion Presence	-.0163
11 Product Champion Influence	-.1728
12 Project Leaders Position	-.1083
13 Project Leaders Power	-.1924
14 Project Leaders Tenure	-.1596
15 Project Leaders Involvement	-.0101
16 Project Members Education	-.1255
17 Project Members Experience	.0137
18 Project Members Tenure	-.3799 **
19 Project Members Involvement	-.0682
20 Representativeness of Interest Groups	-.0336
Organizational Capability - Structure	
21 Team Autonomy	-.3046 *
22 Concurrent Development	--------
23 Turf-Guarding	.3087 *
24 Design-For-Manufacturability	.0952
25 Team Proximity	-.1446
26 Milestone Frequency	.0556
27 Testing Frequency	.6508 **
28 Use of CAD Systems	-.4215 **

$$ᵞ = p<.10$$
$$* = p<.05$$
$$** = p<.01$$

Table 4: List of antecedent factors inputted into the backward-elimination regression procedures and zero-order correlations with concurrent development

5. Discussion

By itself, concurrent development is a non-predictor of cost and quality and a marginal predictor of speed. However, when viewed in the context of other variables active in the innovation process (e.g., strategy, goals, leaders, team proximity, milestones, etc.), concurrent development predicts all three dependent variables at statistically significant levels. These effects, however, are contingent on the other independent variables in the regression equation and are also affected by the correlations of the independent variables (SPSS 1994). They may also suggest the existence of moderated (Baron, Kenny 1986) effects. Thus, these findings provide a more complex, albeit messier answer to the research questions but perhaps a more accurate one.

The comparison of backward-elimination regression models produced several interesting observations. First, it was found that a greater degree of concurrent development *in*creased the speed (i.e., *de*creased the time) of the new product development process ($p<.001$). The beta coefficient for concurrent development ($\beta=.601$) was the highest of the six variables selected for the final regression model, suggesting that its relative importance in predicting speed is high.

Indeed, this finding is consistent with the experiences of many firms experimenting with overlapping activities. Izuchukwe (1992) reports that concurrent engineering practices helped Xerox reduce its time-to-market for new copiers 50 percent, AT&T reduce its time-to-market for a new microprogrammed digital switch by over 40 percent, Boeing reduce its cycle time 40 to 60 percent, and John Deere reduce its new product development time for construction equipment by 60 percent. Hull, Collins, and Likers (1996) report that, when combined with other variables such as design controls and information technology, concurrent engineering processes significantly reduced the time of product development in a research sample of 74 companies. Gilbert (1995) also reports that evidence from the Ford Taurus development project supports the time-saving potential of concurrent engineering.

Second, it was found that a greater degree of concurrent development *in*creased the cost of the new product development process ($p<.01$). The beta coefficient for concurrent development ($\beta=.650$)

was the highest of the eleven variables selected for the final regression model, suggesting that its relative importance in predicting costs is high.

This finding is contrary to Hypothesis 2 insofar as it is consistent with the notion that there is a trade-off in using concurrent development between speed and costs. That is, overlapping activities might speed up the process but it also might cost more to execute (Crawford 1992). Graves (1989) offers four possible explains for this trade-off. First, when steps are overlapped, each task is begun with less information. This results in mistakes and rework, for example on the Xerox 1045 copier project. The greater the overlap, the more tasks are initiated under uncertainty and the greater the subsequent cost penalties. Second, with greater overlap, communication burdens increase and, because of the increased cost of maintaining the more complex communication networks, marginal productivity declines. Third, with greater overlap, there is greater redundancy. That is, more approaches to tasks are undertaken at once, increasing the chances that some will be paid for yet be unsuccessful. Conversely, with more sequential development, the financial risks of failed tasks are limited insofar as only those approaches are tried which precede the discovery of one, which is successful. Fourth, with greater overlap, the network of tasks being completed at any one time becomes increasingly dense. Thus, each successive attempt at further overlap requires the adaptation of a cumulatively greater number of tasks, driving total project costs up.

The opposite effects of concurrent development on speed and on costs may also be explained by examining the relationship between time and money. Several authors (Gupta et al. 1992; Murmann 1994; Vincent 1989) argue that relationship resembles a U-shaped function, in which accelerating development reduces costs up to a point; after that point more expenditures are required to shorten the time to bring products to market (see Figure 3). Shortening development time below the function's minimum (i.e., moving up the "U" to its left) increases costs because of additional coordination expenditures, thereby burning resources because it pushes functions to the limit of organizational capabilities (Vincent, 1989). Alternatively, lengthening development time above the function's minimum (i.e., moving up the "U" to its right) increases costs because of lost learning, reduced motivation, and higher variable expenditures, thereby wasting resources due to dissipated efforts and lapses of attention.

Thus, when firms create an "overly" speedy process through over-lapping activities, costs are likely to increase.

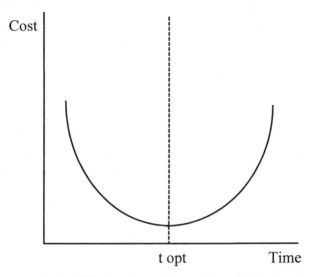

Figure 3: Trade-off between time and money in the new product development process

Moreover, it might be the case that this speed-cost trade-off will be steeper under certain conditions (Graves 1989). One, when there is greater uncertainty in innovation projects (i.e., they are more radical) because greater uncertainty leads to more severe informa-tion gaps. Two, with larger firms because of the greater planning and coordination costs involved in overlapping projects. Three, with less experienced firms because they are more likely to make costly mis-takes in an accelerated process. Four, with larger projects because there is a denser network of tasks subject to greater compression costs. Thus, when these conditions exist, it might be especially unwise to overlap processes from a cost perspective.

Third, it was found that a greater degree of concurrent develop-ment *de*creased the quality of the end product ($p < .05$). The beta coefficient for concurrent development ($\beta = -.281$) was the fourth highest (or fourth lowest) of the seven variables selected for the final regression model, suggesting that its relative importance in predict-ing quality is moderate. This result is contrary to the prediction of Hypothesis 3. Instead, it reinforces the argument that concurrent development, along with similar techniques designed to speed up innovation processes and cut the time it takes to bring a new product to market, may result in more mistakes (i.e., lower quality). This is because, in many cases, concurrent development allows for insufficient time to adapt already commenced stages to the results of market and beta tests (Crawford 1992). These effects of concurrent engineering may be especially detrimental to the quality of radical innovations (i.e., those involving a greater departure from the status quo), which require more information and learning as well as greater care in execution. Indeed, Handfield (1994) found that concurrent development increased the percentage of defects for breakthrough innovations.

Additionally, this finding is consistent with previously cited research which suggests concurrent development is but *one* component of an integrated total quality management (TQM) program primarily, and it is primarily concerned with process issues and the nature of production (Reed et al. 1996). This is in contrast to other components of TQM concerned with market outcome, or the attraction and retention of customers by satisfying their needs better than rival organizations (i.e., quality).

It is also important to understand that concurrent engineering is generally intended for relatively predictable project environments. In turbulent environments, this practice lacks the ability to gather and rapidly respond to new knowledge as the project evolves. Thus Iansiti (1995) proposes that a flexible approach, in which proactive management increases adaptation capabilities, goes beyond traditional concurrent engineering. That is, concurrent engineering implies the joint participation of different functional groups (i.e., joint problem solving) but not necessarily the simultaneous, reciprocal execution of conceptualization and implementation. As a result, concurrent engineering is not (and should not be) intended to react to technological, demographic, and competitive turbulence in the environment. In other words, concurrent engineering may help execute projects quickly (i.e., speed) but not necessarily help projects adapt to the changing environment in line with user demands (i.e., quality).

Overall, concurrent development was only found to predict speed, costs, and quality when viewed within the context of other variables active in new product innovation. Correlations helped to reveal some of the interrelationships underlying this observation. For example, concurrency was associated with greater emphasis placed on speed (versus costs or quality) by top management and a greater reward system orientation for speed. When top management put speed ahead of quality or costs, concurrent development tended to be adopted more frequently. When rewards were dispersed based on speed-based criteria, concurrent processes tended to reflect this direction. This can be seen to reinforce the notion that concurrent development was primarily used as a tool for speed. In a word, firms got precisely what they emphasized and what they rewarded.

Concurrency was associated with lower clarity (i.e., greater ambiguity) surrounding both the project goals and the product concept. These findings are consistent with the idea that overlapping activities greater degrees of concurrent development tends to make processes more fluent and dynamic, and that it also tends to make issues less clear-cut. Therefore, concurrent development might afford development personnel greater flexibility in their activities but at the sacrifice of lower standardization and less control.

Concurrency was associated with more radical innovations. This is consistent with the idea that frame-breaking changes need to be fast to hit the market first to secure pioneering advantages. Alternatively, more incremental changes adapt already accepted standards and to improve on them in some way (e.g., performance-enhancement or price-reduction), thereby necessitating a lesser need for speed. This suggests that use of concurrent development can be thought of as a strategic issue which needs to be fit to the type of product produced as well as the nature of the market engagement.

Concurrency was associated with lower project member tenure. That is, newer project team members were likely to be involved with overlapped processes whereas members who were relatively more entrenched in their organizations tended to be associated with more traditional processes. This is somewhat intuitive insofar as outsiders are typically more receptive to change than insiders. Consequently, this finding may reflect underlying issues related to human-resource management (e.g., people are matched to processes) or socio-political dynamics (e.g., people determine processes based upon personal factors).

Concurrency was associated with lower authority (i.e., less empowerment) of project teams. In other words, the greater the degree of process overlap, the less power was given to teams and the more power was retained by top management. This could be because concurrent processes need greater coordination and oversight than linear, structured processes. Thus, concurrent development may actually remove decision-making responsibility from the team charged with developing the new product. Of course, it could be the case that this finding reveals a poor (versus inevitable) practice.

Perhaps the results would have been different if concurrent development was combined with empowerment. This suggests that there may be better and worse ways of overlapping activities.

Concurrency was associated with greater turfguarding. That is, the greater the degree of overlap, the more project team members tended to experience functional-area based conflicts and hold fast to department norms and objectives. This "description" of practices is contrary to "prescriptive" arguments and again raises the issue of whether there are better and worse ways to overlap activities. To reap the full benefits of interaction and communication, which come with concurrent development, one may be to reduce the perceived threat of overlap and get people to work together productively. The underlying assumption of this threat-based argument is emotional, whereas the reduction of boundaries and specialized domains of activities may cause project members' defenses to rise and guard their turf as to protect any remaining autonomy, power, identity, and the like. As a result, potential synergies from concurrent development may be reduced.

Concurrent development of products was found to be associated with increased frequency of testing. This means that a greater frequency of testing is required in projects under overlapped processes. Since concurrent development does not follow a linear model, one cannot depend on a few tests at the completion of major milestones in the project. In fact, there are many simultaneous discrete paths to be followed for the product development process. This necessitates testing at the completion of each parallel phase or path. Of course, this adds to the costs of the project.

Finally, we observed that concurrent development projects made less use of CAD (computer aided design) systems. Apparently this may seem to be somewhat surprising. Our explanation is that concurrent development processes involve less formal and discrete transfer of information among the various groups of functional organizations. Since the work of the various groups is very much interdependent, use of CAD systems becomes less important. This may also confirm the view that concurrent development has some inefficient practices embedded in it (Hull et al. 1996). Nevertheless, it raises important questions regarding the efficiencies of alternative communication structures and the positive or negative synergies that arise from the combination of multiple project management techniques.

6. Concluding comments

The mixed results from the regression analyses imply that firms adopt more objective views of concurrent development that consider its potential benefits (e.g., speed) as well as its potential liabilities or trade-offs (e.g., costs and quality). The findings also suggest the question of whether there are better and worse ways of developing projects concurrently. For example, combining concurrent development with a strong emphasis on speed by top management or a narrow silo-orientations among functional areas may produce results quite different than if concurrent development was combined with clear product concepts or strong champion involvement.

Consequently, these results suggest that R&D managers refrain from shotgun approaches to concurrent development and instead, with potential trade-offs in mind, match their development strategy (i.e., whether to use concurrent development and how to use concurrent development) to their project objectives. Thus, concurrent development does not appear to be a panacea and should instead be viewed as a management technique having complex consequences. *Simply, there may be trade-offs in using concurrent practices (i.e., faster speed but higher costs and lower quality) and these trade-offs may be a function of how concurrent development is used (i.e., in combination with what other development techniques).*

It is important to keep in mind that the strength of these implications are conditioned by the low correlations between concurrent development and project outcomes. Thus future research and analysis needs to investigate the complex connections between concurrent development and project outcomes. For example, radicalness may moderate the relationship between concurrent development and speed (Kessler, Chakrabarti 1998). Moreover, these are only suggestions borne from one study of large, U.S.-based firms considering product innovations. Future research considering firms of different sizes, firms in different national contexts, and innovations of different types (e.g., process) (Gopalakrishnan et al. 1998) is needed to clarify important boundary conditions and the subsequent generalizability of the findings.

Patterns in high-impact innovation

George F. Farris

Patterns in high-impact innovation

George F. Farris

1. Introduction

All innovations are not equal. Some have had extraordinary impact on economic growth, job creation or the standard of living, while others have had much less impact. Similarly, some innovations have been of great usefulness or value to the organizations in which they occurred, while others have been less useful in helping them to carry out their responsibilities. A few individuals have been major contributors to high-impact innovations, but most others have done work of substantially less impact or value. It is important to understand what distinguishes these high-impact key contributors from very good scientists and engineers. The present paper addresses this topic.

Although high-impact innovation is a topic of great practical importance, it has only recently begun to receive the attention it deserves. *Research-Technology Management*, for example, recently published articles on the topic describing "breakthrough innovation" at Hewlett Packard (Barnholt 1997) and "discontinuous innovation" at DuPont (Norling, Statz 1998), and scholars from RPI are beginning to publish results of their major study of breakthrough innovation (Lynn, Mazzuca, Morone, Paulson 1998; Rice, O'Connor, Peters, Morone 1998).

The role of key individual contributors has not been a focus of the work to date, however, probably because it is so difficult to study systematically. Key contributors to high-impact innovations are few in number by definition, and they are only rarely available to participate in studies. Instead, most research on individual scientists and engineers has studied the entire range of performance, from marginal to average to very good to excellent in order to determine correlates of performance.

Pelz and Andrews' (1976) classic study of scientists in organizations is typical of this approach. They surveyed 1311 scientists and engineers from eleven laboratories in industry, universities and government. A questionnaire was given to the participants in the study, senior scientists and managers rated their performance and the questionnaire responses were then correlated with the performance measures. The study identified many characteristics of "productive climates" for research and development, including freedom, communication patterns, diversity of work activities, coordination and motivations. In general, lower performers scored lower on these factors than those rated higher.

These findings are interesting and important, but they do not directly address high-impact innovation. Consider communications, for example: lower performers talked to fewer people than did high performers. This finding could have occurred because low performers talked to fewer people than average performers or because average performers talked to fewer people than high performers. The finding does not tell us about differences between high performers and outstanding performers, say those in the 90th percentile and above. The outstanding performers may have talked to more people, fewer people, or the same number of people as the high performers. And it is these outstanding performers who are apt to be key contributors to high-impact innovation.

It is the purpose of this paper to identify characteristics of the small number of key individuals who make major contributions to high-impact innovations. The paper reports on two studies of this problem. First, interviews were held with 28 recipients of the National Medal of Technology, an award given by the president of the United States to recognize those "whose inventions have contributed dramatically to an improved standard of living, job creation, and economic growth". Patterns of characteristics were identified in these interviews. Second, these patterns were tested by comparing outstanding contributors from 24 industrial laboratories with colleagues rated as high, but not outstanding, contributors. Finally, based on the results of the two studies, tentative conclusions were drawn about common characteristics of key contributors to high-impact innovation.

2. The interview study

Interviews were held with 28 recipients of the National Medal of Technology, representing about one third of the individuals who have received the Medal since it was first presented in 1985. The Medal, inspired by the Nobel Prize, is awarded annually by the president of the United States in a ceremony at the White House. The recipients are recommended by a nominating committee which reviews the nominations received each year, and the president makes the final decision. Typically about 90 nominations are received and about five medals are awarded each year. Recipients of the Medal who were interviewed and their contributions are listed in Figure 1. The interview guide is shown in Figure 2. The interviews lasted about two hours and were tape recorded. They were held at the medalist's office or home.

In each interview several factors were identified which appeared to contribute to the person's achievement. Identifying factors which had affected most of the medalists proved much more difficult, however. Some factors were unique to the individual or held in common by only a small number of medalists. For example, two individuals described themselves as voracious readers from poor families who made extensive use of public libraries. Other factors were mentioned by a majority of the medalists. For example, many of them identified a source of support or security during childhood or adulthood, usually a parent or spouse.

Identification of common factors was made more difficult by an insight that the National Medal of Technology had apparently been awarded to three different kinds of high-impact contributors: technological entrepreneurs, who either founded or greatly expanded technology-based companies; technology champions in established firms, who led major development efforts, and creative technologists, who developed high-impact technology but relied heavily on others for its promotion. The classification of the medalists interviewed is shown in Figure 3. It is likely that all three types of contributors share some characteristics, but that each type has particular characteristics which distinguish it from the other types. Unfortunately, it has not been possible so far to identify these distinguishing characteristics.

Person	Company	Achievement
1. Walt Robb	GE	Led MRI, Cat Scan development
2. John Mayo	Bell Labs	CEO, Fiber optics, switches
3. George Kozmetsky	U. Texas	Co-founder of Teledyne
4. Gordon Moore	Intel	Co-founder and CEO
5. Gordon Bell	Digital Equip	Architect of PDPs & Vaxes
6. Bob Galvin	Motorola	CEO
7. Joel Engel	Bell Labs	Cellular phone (now @ Ameritech)
8. Del Meyer	Amoco	Polyester production process
9. Dick Frenkiel	Bell Labs	Cellular phone (now @ Rutgers)
10. Dave Duke	Corning	Vice Chair/CTO (co. award)
11. Don Stookey	Corning	Photochromic glass, Corningware
12. Irwin Jacobs	Qualcomm	Founder/CEO. CDMA/wireless tech.
13. Allen Puckett	Hughes	CEO. Geostationary satellite
14. Gordon Binder	Amgen	CEO. Biotechnology (co. award)
15. Joe Sutter	Boeing	727,737,757,767, "Father of 747"
16. Ken Iverson	Nucor	Minimills. Revitalized steel industry
17. Marinus Los	Am.Cyanamid	Environmentally friendly herbicides
18. David Thompson	Orbital Sci	Founder/CEO. (Pegasus Team)
19. George Levitt	DuPont	Environmentally friendly herbicides
20. David Pall	Pall Corp.	Filtration products - medical, oil
21. Bill Joyce	Union Carbide	CEO; UNIPOL--polyethylene
22. Steve Wozniak	Apple	Co-founder
23. Ken Olsen	Digital Equip	Founder, CEO
24. Roy Vagelos	Merck	CEO (company award)
25. Praveen Chaudhari	IBM	Magneto-optical disk drives (team award)
26. Jerry Cuomo	IBM	Magneto-optical disk drives (team award)
27. Dick Gambino	IBM	Magneto-optical disk drives (team award)
28. Ray Dolby	Dolby Labs	Audio enhancement technology

Figure 1: National Medal of Technology medalists interviewed
(listed in order interviewed)

1. **Your achievement.** I've read your nomination for the Medal. The citation says "...". Could you give me an idea of some of the things you did to earn that citation?
   - What were some of the key events in the process?
   - What facilitated the process?
   - Did you have to overcome any obstacles? What were they - technical, organizational, financial?
   - Were there any crises?
   - About how many hours a week did you work?
   - Are there any <u>stories</u> related to the achievement which would "stick in peoples' heads?"
   - Would you like to comment on the impact of your work?

2. **Your career.** I would like you to reflect on your career, starting with school and going up to the time of your achievement.
   - Can you recall how you first became interested in your technical field? - At what age?
   - Were you influenced by any role models?
   - When you began to work full time, what "kept you going?" How were you rewarded for your accomplishments? How did these rewards influence you?
   - Did you have any mentors during your career?
   - Is there anything else in your working environment or your life outside work which enabled you to achieve what you did?

3. **Risk taking.** Technical work and technical careers frequently require taking risks - personal, financial, or technical. Can you think of one or more risks you took in your work or career? What were the outcomes? Were other people involved? What role did they play in your decisions to take risks?

4. **Advice.** What advice would you give to someone today who would like to receive a National Medal of Technology? Let's think about two people:
   - A scientist or engineer early in his or her career
   - A student who may be interested in pursuing a technical career

5. **Other comments.** Are there any comments or suggestions you would like to add to make this project as useful as possible?

Figure 2: Interview guide

Type	Name
Technological Entrepreneurs	Dolby, Galvin, Iverson, Jacobs, Kozmetsky, Moore, Olsen, Pall, Thompson
Technology Champions in Established Firms	Engel, Joyce, Mayo, Puckett, Robb, Sutter
Creative Technologists	Bell, Chaudhari-Cuomo-Gambino team, Frenkiel, Levitt, Los, Meyer, Stookey, Wozniak
Company Awards	Amgen (Binder), Corning (Duke), Merck (Vagelos)

Figure 3: Types of medalists

Several factors were identified which were common to most of the medalists. They are listed in Figure 4 and summarized below. The list is not exhaustive, but it includes many of the themes emerging from the interviews.

- Persistence - tenacity approaching obstinacy

- Calculated risks accompanied by security

- Hard work accompanied by fun doing it

- Early success common, but some "late bloomers"

- Childhood spark and restless impatience

- Riding the waves of technology and luck

Figure 4: Factors common to medalists

Persistence - tenacity approaching obstinacy. The first comment made by Joel Engel, who received the Medal along with Richard Frenkiel for developing the cellular telephone system, was, "This should be a medal for *obstinacy*". He described years of convincing supervisors and management at Bell Laboratories and AT&T that the concept was technically and economically viable, and then AT&T spent years getting the government approval needed to implement the system. Many other medalists told similar stories of early failures or slow progress followed by ultimate high-impact success.

Calculated risks accompanied by security. In most instances the medalists knew they were contending with difficult odds as they worked on their achievements, but they calculated that they could overcome these odds and succeed. Moreover, most were quite self-confident, based on sources of support in childhood and/or adulthood, and many of them had been led to expect "soft landings" if their efforts failed. Walter Robb of General Electric, for example, was given eighteen months to turn around the company's poorly performing medical equipment business. If he failed, the business was going to be sold and he would have the choice of going with the business or returning to an equivalent position at General Electric.

Hard work accompanied by fun doing it. Most of the medalists reported that they worked intensely on the activities which led to their receiving the Medal. Often they reported working long hours, but some medalists emphasized that they took their vacations. This hard work was not drudgery, however. Many medalists remarked on the fun that work was for them, both the work itself and the congeniality with colleagues.

Early success common, but some late bloomers. The first year on the job is often instrumental in launching a successful career, and challenging tasks during that year are critical. This was the case for many of the medalists who worked in corporations, but there were notable exceptions. George Levitt of DuPont, developer of an environmentally friendly herbicide, for example, described himself as a "journeyman chemist" until he succeeded in developing the herbicide when he was in his fifties.

Childhood spark and restless impatience. Technical careers are not synonymous with single technical jobs. Some scientists and engineers develop technical interests early in life, and some change employers later in their careers. Many of the medalists became excited about technology very early in life. Joseph Sutter, the "Father of the 747", fondly recalled riding his bicycle as a child to Boeing Field in Seattle to watch test flights. Articles have been written about Gordon Bell, the "boy electrician" who at age twelve led appliance repair crews in a small town in Missouri. Gordon Bell received the Medal for his contribution as architect of Digital Equipment's VAX computers. On the other hand, some Medalists developed their technical interests relatively late in life. Delbert Meyer, who developed a polyester manufacturing process for Amoco, expected to farm his family's land in Iowa after he returned from military service. Instead, he received a government scholarship, attended college, discovered the field of chemistry and changed careers.

Many Medalists stayed with their employers throughout their careers, but several quit to start their own companies or join other companies. For example, David Packard left General Electric to form Hewlett Packard, Steve Wozniak left Hewlett Packard to form Apple Computer, and Gordon Bell recently left Digital Equipment to join Microsoft.

Riding the waves of technology and luck. Most of the Medalists were quite modest in taking credit for their achievements, attributing them to colleague support, organizational support, availability of technology, or just plain good fortune. "Riding the waves of technology" was a recurrent theme, expressed, for example, by Gordon Moore, co-founder of Intel and John Mayo, former CEO of Bell Laboratories. Others referred to being in the right place at the right time or being born at the right time.

These themes emerged from interviews with twenty-eight individuals recognized for the impact of their achievements. Most of them were common to the majority of the medalists, but exceptions did occur as noted. It is important to determine whether

(1) these themes would emerge in a larger sample of high-impact innovators and

(2) whether these themes would distinguish outstanding innovators from high, but not outstanding, contributors.

The questionnaire study provides some initial information on these issues.

3. The questionnaire study

As part of an earlier study, an extensive paper-and-pencil question-naire was completed by 3,163 scientists and engineers working in research and development laboratories of 24 companies in the United States. The sections of the questionnaire described aspects of the work environment, career, relations with others, the supervisor, work and family life and personal background information. Managers and senior technical staff rated the performance of the individuals who completed the questionnaire on dimensions including "the extent to which each person's work has been *useful or valuable* in helping your R&D organization to carry out its responsibilities". The eleven-point scale ranged from 0 to 99, and the rater was asked to compare the individual to the other technical staff in the company and indicate a percentile score. An average of three judges rated each individual's performance.

A total of 2,573 individuals were rated on the value of their work. Of those, 183 (7.1% of those rated) were rated as doing work whose value was greater than that of 90% of their colleagues. These individuals were designated "outstanding" contributors. Another 1,069 (43.2% of those rated) individuals were rated as doing better than 66.67% of their colleagues but less well than the top 10%. These were designated "high" contributors. The outstanding and high contributors were compared in the analyses which follow using t-tests for independent samples.

Preliminary analyses indicated no significant differences between the outstanding and the high contributors in age, country of birth (U.S. or abroad), or percent holding the Ph.D. There were slightly fewer women among the outstanding performers (17.1% vs. 21.2%). Correlations among the predictors were quite low.

Among the frequent themes common to the interviews with the twenty-eight medalists were persistence, risk taking, security, hard work, fun and restless impatience. These themes were selected for investigation in the questionnaire study because the questionnaire included items which appeared to measure them. The questions and the results of the analysis are indicated in Figure 5.

Factor	Contributor:			
	Out-standing	High	t	p*
Persistence				
Like to probe deeply and thoroughly	5.48	5.29	1.80	.036
Self-rating of energy	7.63	7.36	1.90	.029
Risk taking				
Consequences of taking a risk which fails (high number is more benign)	3.68	3.67	0.05	ns.
Security				
Opportunity to have stability of employment	4.58	4.65	-0.57	ns.
Hesitancy to share ideas freely in work group	2.64	2.60	0.31	ns.
Hesitancy to share ideas freely with other groups	3.48	3.50	-0.13	ns.
Hard work				
Working hours at regular company work site	44.03	43.08	1.82	.034
Total working hours per week	50.92	50.01	1.30	ns.
Involvement in work	4.94	4.87	0.99	ns.
Fun				
Provision for congenial colleagues as co-workers	4.53	4.59	-0.64	ns.
Opportunity to enjoy work	4.31	4.46	-1.16	ns.
Restless impatience				
Likelihood of starting own company within 10 years	2.13	2.11	0.13	ns.

*One-tailed test

Figure 5: Comparison of outstanding and high contributors on predicted factors

Of the themes tested, only *persistence* and *hard work* significantly (one-tailed tests) distinguish the outstanding from the high contributors. The outstanding contributors were more apt to "like to probe deeply and thoroughly into problems" and to rate themselves as higher in "energy (hard work, drive, meeting schedules; large

High-impact innovation   315

output)". They also reported that they worked more hours per week at their regular company work sites.

Given these disappointing results, a second analysis was conducted. The outstanding contributors were compared with the high contributors on nearly all the remaining items in the questionnaire, using two-tailed tests of significance. This unorthodox approach was deemed useful to try to determine tentatively whether better questions were available to indicate themes from the interviews and whether new themes emerged. Figure 6 shows the questions for which there were significant differences.

Factor	Contributor:			
	Out-standing	High	t	p*
Importance of clear objectives	5.48	5.28	2.05	.041
Number of short (<1 year) assignments abroad	1.12	0.76	2.14	.032
Number of foreign-born colleagues in group	2.21	1.87	1.95	.050
Group members have many different opinions	4.42	4.19	2.05	.040
Group members have similar work styles	2.79	3.06	-2.53	.011
Group has been working together a long time	3.80	3.49	2.15	.032
Number of men in group	2.96	3.07	2.20	.028
Group incorporates R&D concerns before decision	4.48	4.02	2.82	.005
Ease of getting cooperation from supervisor	5.41	5.19	2.04	.042
Self-rated creative ability	7.30	6.94	2.17	.030
Confidence others have in your abilities	5.79	5.64	2.02	.044
Patents in the last 5 years	2.24	1.47	2.49	.013
Have a problem finding time to care for myself	3.55	4.20	-2.39	.017
Want job with sufficient time for personal life	5.71	5.89	-1.98	.047
Flexible hours policies help me perform better	5.17	5.47	-2.55	.011
Opportunities to build outside reputation	2.58	2.90	-2.62	.009
Career was influenced by the example of a successful more experienced professional	3.31	3.56	-1.95	.050
Likely to move to different laboratory in company	2.89	3.21	-2.27	.023

*Two-tailed test

Figure 6: Comparison of outstanding and high performers on other factors

A number of differences emerged from this analysis. Outstanding contributors, as compared to high contributors, were higher in the importance they placed on having clear objectives and in their exposure to sources of a greater variety of ideas, especially from outside the United States. They had more experience working abroad, had more foreign-born colleagues in their work groups, and were more apt to consider many different opinions before reaching a decision. Their groups had also been together longer and included more men. They found it easier to get cooperation from their supervisors, rated their creative ability higher and felt that others had greater confidence in their abilities. They reported that they had produced more patents during the last five years. They also reported that cross-functional groups to which they belonged were more apt to incorporate the concerns of R&D before they made any important decisions in other functional areas.

High contributors, as compared to outstanding contributors, appeared to have greater difficulty balancing work and personal life, greater orientation outside their organizations, and greater influence on their careers from role models. They reported that they had greater difficulty finding time to take care of themselves, that personal time was more important to them, that their companies' flexible hours policies were more apt to help them to perform better, that they had greater opportunities to build their reputations outside the organization, that they had been influenced more by the examples of the success of experienced professionals, that members of their groups had more similar work styles, and that they were more apt to transfer to another laboratory within the company.

4. Discussion

A summary of the results of the questionnaire study appears in Figure 7. Many of the themes emerging from the questionnaire analysis, in hindsight, were consistent with things said in the interviews. Clear objectives were implicit in the achievements of most, if not all, of the medalists. Many of them had been exposed to a variety of sources of ideas. Steve Wozniak, for example, was active in a computer club before co-founding Apple Computer, and Ray Dolby studied in England and started his business there before returning to the United States. The medalists were quite adept at balancing work

---

Outstanding Contributors Are Higher in

- Persistence - probing deeply, energy, working hours
- Desire for clear goals
- Exposure to differing opinions and ideas

High Contributors Are Higher in

- Concerns for balancing work and personal life
- Similar styles in work group
- Being influenced by a role model
- Opportunity to build reputation outside the company

Figure 7: Summary of questionnaire study results

and family life. They were focused on their own organizations rather than their outside reputations, (except in those instances where they changed jobs). Not one of them acknowledged having a role model.

Before drawing conclusions from this study it should be noted that they must be regarded as tentative. Twenty-eight two-hour interviews with medalists provide useful, but incomplete information. Analysis of questionnaire data collected for a different purpose can only begin to identify themes common to outstanding contributors and medalists and to identify possible differences between outstanding and high contributors.

Let us try to draw meaningful conclusions, recognizing that they are quite tentative. Combining the findings of the interview and questionnaire studies, what patterns emerge to distinguish outstanding from high performers? Figure 8 summarizes some key patterns. Outstanding contributors are focused on clear goals. They are persistent, energetic and hard working, probing deeply and thoroughly. They enjoy their work. They are exposed to a diversity of ideas through multiple contacts with people holding different opinions, for example through cross-functional or international work.

- High Contribution Requires Security and Challenge
- Outstanding Contribution Requires Focus and Diversity
- Facilitators of High-Impact Innovation Include:
  - Persistence
    = Energy
    = Hard work
    = Probing deeply and thoroughly
  - Exposure to diversity of ideas
    = Cross-functional work
    = Heterogeneous groups
    = International exposure
- Barriers to High-Impact Innovation Include
  - Problems balancing work and personal life
  - Greater emphasis on building reputation
  - Greater influence by role models
  - Greater similarity of work styles in group

Figure 8: Patterns in high-impact innovation:
     Tentative conclusions based on both interview and questionnaire studies

High contributors, on the other hand, have greater difficulty maintaining focus, since they are more involved with or concerned with balancing work and family life. They direct more of their attention outside their laboratories, maintaining their outside reputations and sometimes transferring to other laboratories, and they have been influenced more by role models.

Donald Pelz (Pelz, Andrews 1976) argues that "creative tensions" between sources of security and challenge produce productive climates for scientists and engineers. For example, effective scientists were self-reliant but interacted vigorously with colleagues. Here the self-reliance is a source of security and the ideas from the vigorous interaction are a source of challenge. Pelz and Andrews were concerned with finding correlates of performance among the entire technical staffs of R&D laboratories.

In the present study the concern is different: to identify factors which distinguish high performers from those making truly out-standing, high-impact contributions. Based on the tentative results of this study, it seems that security and challenge are necessary but not sufficient. Here the important creative tension appears to be between *focus* and *diversity* - between an energetic persistence and exposure to a variety of ideas which may help solve the problem. This focus is facilitated by the absence of barriers which occur more frequently for less outstanding individuals, especially problems in balancing work and personal time.

These tentative conclusions should serve as guidelines for future research on high-impact innovation. They need to be confirmed in interview studies with more interviews and in comparative studies, using interviews or questionnaires, specifically *designed* to compare high contributors with outstanding contributors. Such studies should provide a greater understanding of the high-impact innovations which contribute so "dramatically to an improved standard of living, job creation, and economic growth".

Acknowledgement

Paper prepared for presentation at Conference "Dynamics of Inno-vation Processes," Hamburg, July 16-17, 1998. The author gratefully acknowledges the collaboration of Nancy DiTomaso of Rutgers University and Rene Cordero of the New Jersey Institute of Tech-nology in the questionnaire study, the support of the United States Department of Commerce in the interview study, the support of the Alfred P. Sloan Foundation and the Center for Innovation Manage-ment Studies at Lehigh University in the questionnaire study, and the support of the Technology Management Research Center of Rut-gers University in both studies.

Literature

Abernathy, W.J., Utterback, J.M., Patterns of industrial innovation, Technology Review, June - July, 1978, pp. 40-47.

Achilladelis, B., The dynamics of technological innovation: The sector of anti-bacterial medicines, Research Policy, vol. 22, 1993, pp. 279-308.

Ackoff, R.L., Creating the corporate future, New York: Wiley 1981.

Albach, H., Culture and Technical Innovation - A Cross-Cultural Analysis and Policy Recommendations, Berlin/New York: de Gruyter 1993/94.

Albert, M.B., Avery, D., Narin, F., McAllister, P., Direct validation of citation counts as indicators of industrially important patents, Research Policy, vol. 20, 1991, pp. 251-259.

Aldenderfer, M.S., Blashfield, R.K., Cluster analysis, California et al.: Sage University Paper 1984.

Aldrich, H., Sasaki, T., R&D Consortia in the United States and Japan, Research Policy, vol. 24, 1995, pp. 301-316.

Amelingmeyer, J., Gerhard, B., Specht, G., The Influence of External Participation on the Success of Innovation Processes in Business-to-Business Markets, in: Mazet, F., Salle, R., Valla, J.-P., Interactions, Relationships and Networks. Proceedings of the 13th IMP Conference, Work in Progress Papers, vol. 1, Lyon: Ecole Superieur de Commerce 1997, pp. 5-19.

Ancona, D.G., Caldwell, D.F., Improving the performance of new product teams, Research Technolog Management, vol. 33, 1990, pp. 25-29.

Ancona, D.G., Caldwell, D.F., Bridging the Boundary: External Activity and Performance in Organizational Teams, ASQ, vol. 37, 1992, pp. 634-665.

Anderson, P., Tushman, M.L., Technological discontinuities and dominant designs: A cyclical model of technological change, Administrative Science Quarterly, vol. 35, 1990, pp. 604-633.

Armstrong, J., Reinventing Research at IBM, in: Rosenbloom, R.S., Spencer, W.J. (Eds.), Engines of Innovation. U.S. Industrial Research at the End of an Era, Cambridge, MA: MIT Press 1996, pp. 151-154.

Arrow, K.J., The Limits of Organization, New York: Norton 1974.

Ashton, W.B., Kinzey, B.R., Gunn Jr., M.E., A structured approach for monitoring science and technology developments, International Journal of Technology Management, vol. 6, 1991, pp. 91-111.

Ashton, W.B., Sen, R.K., Using patent information in technology business planning-I, Research Technology Management, November - December, 1988, pp. 42-46.

Bacher, J., Clusteranalyse: anwendungsorientierte Einführung, München et al.: Oldenbourg 1994.

Bachofner, M., Vertrauen und Verbundenheit zum Start europäischer kooperativer F&E-Projekte - eine empirische Analyse anhand von ESPRIT-Projekten, Dipl. Thesis University of Karlsruhe 1996.

Backhaus, K., Erichson, B., Plinke, W., Weiber, R., Multivariate Analysemethoden: eine anwendungsorientierte Einführung, 6th edition, Berlin et al.: Springer 1990.

Badawy, M.K., What we've learned: Managing human resources, Research Technology Management, vol. 31(5), 1988, pp. 19-35.

Baker, M.J., McTavish, R., Product policy and management, New York: Macmillan 1976.

Bandura, A.A., Social Foundations of Thought and Action, Englewood Cliffs, NJ: Prentice-Hall 1986.

Bandura, A.A., Organizational applications of social cognitive theory, Australian Journal of Management, vol. 13(2), 1988, pp. 275-302.

Bantel, K.A., Jackson, S.E., Top management and innovations in banking: Does the composition of the top management team make a difference, Strategic Management Journal, vol. 10(S), 1989, pp. 107-124.

Barker, K., Dale, A., Georghiou, L., Management of Collaboration in EUREKA-Projects: Experiences of UK Participants, Technology Analyses & Strategic Management, vol. 8, 1996, pp. 467-482.

Barley, S.R., Freeman, J., Hybels, R.C., Strategic alliances in commercial biotechnology, in: Nohria, N., Eccles, R.G. (Eds.), Networks and organizations, Boston: Harvard Business School Press 1992, pp. 311-347.

Barnard, C., The Functions of the Executive, Cambridge, MA: Harvard University Press 1968.

Barney, J., Hansen, M.H., Trustworthiness as a source of competitive advantage, Strategic Management Journal, vol. 15 (Winter Special Issue), 1994, pp. 175-190.

Barnholt, E.W., Fostering Growth with Breakthrough Innovation, Research Technology Management, vol. 40(2), 1998, pp. 12-16.

Baron, R.M., Kenny, D.A., The moderator-mediator variable distinction in social psychologioal research: Conceptual, strategic, and statistical considerations, Journal of Personality and Social Psychology, vol. 51, 1986, pp.1173-1182.

Barras, R., Towards a theory of innovation in services, Research Policy,vol. 15, 1986, pp. 161-173.

Barras, R., Interactive innovation in financial and business services: The vanguard of the service revolution, Research Policy, vol. 19, 1990, pp. 215-237.

Basberg, B.L., Foreign patenting in the U.S. as a technology indicator. The case of Norway, Research Policy, vol. 12, 1983, pp. 227-237.

Basberg, B.L., Patents and the measurement of technological change: A survey of the literature, Research Policy, vol. 16, 1987, pp. 131-141.

Becker, C., Kooperation als FuE-Strategie? Ergebnisse einer Unternehmensbefragung, in: Gesellschaft für Innovationsforschung und Beratung, VDI/VDE-IT (Eds.), Kooperation als FuE-Strategie in der Mikrosystemtechnik? Berlin/Teltow 1994, pp 16-104 and pp. 172-191.

Becker, C., Zur Kooperationsintensität von ostdeutschen Unternehmen: Neuere empirische Evidenz aus dem Technologiebereich "Mikrosystemtechnik", in: Fritsch, M. (Ed.), Potentiale für einen "Aufschwung Ost". Wirtschaftsentwicklung und Innovationstransfer in den Neuen Bundesländern, Berlin: Edition Sigma 1995, pp. 201-229.

Benkenstein, M., Modelle technologischer Entwicklungen als Grundlage für das Technologiemanagement, Die Betriebswirtschaft, vol. 49, 1989, pp. 497-512.

Bettis, R.A., Hitt, M.A., The new competitive landscape, Strategic Management Journal, vol. 16 (Summer Special Issue), 1995, pp. 7-20.

Bieg, H., Kann der Bankprüfer die Bonität gewerblicher Bankkreditnehmer beurteilen?, Zeitschrift für betriebswirtschaftliche Forschung, vol. 36, 1984, pp. 495-512.

Bierly, P., Chakrabarti, A.K., Generic knowledge strategies in the US pharmaceutical industry, Strategic Management Journal, vol. 17 (Winter Special Issue), 1996a, pp. 123-135.

Bierly, P., Chakrabarti, A.K., Technological learning, strategic flexibility, and new product development in the pharmaceutical industry, IEEE Transactions on Engineering Management, vol. 43(4), 1996b, pp. 368-380.

Bierly, P., Chakrabarti, A.K., Dynamic knowledge strategies and industry fusion, International Journal of Technology Management, 1998, forthcoming.

Bierly, P., Kessler, E., Governance of Interorganizational Partnerships: A Comparison of U.S., European and Japanese Alliances in the Pharmaceutical Industry, in: Hitt, M., Ricart, J.E., Nixon, R.D. (Eds.), Managing Strategically in an Interconnected World, Chichester: John Wiley & Co 1998.

Bingmann, H., Antiblockiersystem und Benzineinspritzung, in: Albach, H., Culture and Technical Innovation. A Cross Cultural Analysis and Policy Recommendations, Berlin, New York: de Gruyter 1994, pp. 736-821.

Birnbaum-More, P.H., Competing with technology in 8-bit microprocessors, Journal of High Technology Management Research, vol. 1, 1990, pp. 1-13.

Bitzer, B., Poppe, P., Strategisches Innovationsmanagement - Phasenspezifische Identifikation innerbetrieblicher Innovationshemmnisse, in: Betriebswirtschaftliche Forschung und Praxis, vol. 45, 1993, pp. 309-324.

Bitzer, B., Innovationshemmnisse im Unternehmen, Wiesbaden: Gabler 1990.

Björkman, I., Lindell, M., Salenius, B.M., Sevón, G., Währn, N., Creation and Change of Commitment to a Collaborative R&D project. The Case of Finnish EUREKA Projects, Working Paper No. 219, Helsinki: Swedish School of Economics and Business 1991.

Blackburn, J.D., Time based competition: White collar activities, Business Horizons, vol. 35(4), 1992, pp. 96-101.

Böhnisch, W., Personale Widerstände bei der Durchsetzung von Innovationen, Stuttgart: Poeschel 1979.

Boeker, W., Strategic Change: The Effects of Founding and History, Academy of Management Journal, vol. 32, 1989, pp. 489-515.

Bound, J., Cummins, C., Griliches, Z., Hall, B.H., Jaffe, A., Who does R&D and who patents?, in: Griliches, Z. (Ed.), R&D, Patents and Productivity, Chicago: University of Chicago Press 1984, pp. 21-54.

Bower, D.J., Keogh, W., Changing patterns of innovations in a process dominated industry, International Journal of Technology Management, vol. 12, 1996, pp. 209-220.

Breuer, R.-E., Bilanzanalyse der Kreditinstitute, Zeitschrift für betriebswirtschaftliche Forschung, special issue, vol. 29, 1991, pp. 151-155.

Bridenbaugh, P.R., The Future of Industrial R&D, or, Postcards from the Edge of the Abyss, in: Rosenbloom, R.S., Spencer, W.J. (Eds)., Engines of innovation. U.S. Industrial Research at the End of an Era, Cambridge, MA: MIT Press 1996, pp. 155-163.

Brockhoff, K., Abstimmungsprobleme von Marketing und Technologiepolitik, Die Betriebswirtschaft, vol. 45(6), 1985, pp. 623-632.

Brockhoff, K., Funktionsbereichsstrategien, Wettbewerbsvorteile und Bewertungskriterien, Zeitschrift für Betriebswirtschaft, vol. 60, 1990, pp. 451-472.

Brockhoff, K., Instruments for patent data analyses in business firms, Technovation, vol. 12(1), 1992, pp. 41-58.

Brockhoff, K., R&D Co-operation between Firms - A Perceived Transaction Cost Perspective, Management Science, vol. 38, 1992, pp. 514-524.

Brockhoff, K., Produktpolitik, 3rd edition, Stuttgart: Gustav Fischer 1993.

Brockhoff, K., Die Bedeutung der betriebswirtschaftlichen Ausbildung für den Standort Deutschland. Betriebswirtschaftslehre und der Standort Deutschland, Zeitschrift für Betriebswirtschaft, Special Issue 1, 1996, pp. 1-6.

Brockhoff, K., Technology Management in the Company of the Future, Technology Analysis & Strategic Management, vol. 8, 1996, pp. 175-189.

Brockhoff, K., Steuerung der Forschung durch abgestimmten Potentialaufbau, Zeitschrift für Betriebswirtschaft, vol. 67, 1997, pp. 453-470.

Brockhoff, K., Industrial Research for Future Competitiveness, Berlin et al.: Springer Verlag 1997.

Brockhoff, K., Forschung und Entwicklung, Planung und Kontrolle, 5. edition, München: Oldenbourg 1998.

Brockhoff, K., Leistungen der Betriebswirtschaftslehre für Wirtschaft und Gesellschaft, unpublished manuscript, University of Kiel 1998.

Brockhoff, K., Chakrabarti, A.K., R&D/Marketing linkage and innovation strategy: Some West German experience, IEEE Transactions on Engineering Management, vol. 35(3), 1988, pp.167-174.

Brockhoff, K., Chakrabarti, A.K., Hauschildt, J., Pearson, A.W., Managing Interfaces, in: Gaynor, G.H. (Ed.), Handbook of Technology Management, New York: McGraw Hill 1996.

Brockhoff, K., Gupta, A.K., Rotering, C., Inter-Firm R&D Cooperations in Germany, Technovation, vol. 11, 1991, pp. 219-229.

Brown, W.B., Karagozoglu, N., Leading the way to faster new product development, Academy of Management Executive, vol. 7(1), 1993, pp. 36-47.

Bruce, M., Leverick, F., Littler, D., Wilson, D., Success Factors for Collaborative Product Development: A Study of Suppliers of Information and Communication Technology, R&D Management, vol. 25, 1995, pp. 33-44.

Bürgel, H.D., Haller, C., Binder, M., F&E-Management, München: Vahlen 1996.

Burgelman, R.A., A process model of internal corporate venturing in the diversified major firm, Administrative Science Quarterly, vol. 28, 1983, pp. 223-244.

Burgoon, J.K., Pfau, M., Parrott, R., Birk, T., Cooker, R., Burgoon, M., Relational communication satisfaction, compliance-gaining strategies, and compliance in communication between physicians and patients, Communication Monographs, vol. 54, 1987, pp. 307-324.

Burns, T., Stalker, G.M., The management of innovation, London: Tavistock 1961.

Butler, J.K., Jr., Toward Understanding and Measuring Conditions of Trust: Evolution of a Conditions of Trust Inventory, Journal of Management, vol. 17(3), 1991, pp. 643-663.

Buzzacchi, L., Colombo, M.G., Mariotti, S., Technological regimes and innovation in services: The case of the Italian banking industry, Research Policy, vol. 24, 1995, pp. 151-168.

Buzzell, R.D., Gale, B.T., The PIMS Principles: Linking Strategy to Performance, New York, London: The Free Press 1987.

Capon, N., Farley, J.U., Hoenig, S., Determinants of financial performance: A meta-analysis, Management Science, vol. 36(10), 1990, pp. 1143-1159.

Carpenter, M.P., Narin, F., Woolf, P., Citation rates to technologically important patents, World Patent Information, vol. 3(4), 1981, pp. 160-163.

Chacko, H.E., Methods of upward influence, motivational needs, and administrators' perceptions of their supervisors' leadership styles, Group & Organization Studies, vol. 15(3), 1990, pp. 253-265.

Chaganti, R., Damanpour, F., Institutional Ownership, Capital Structure, and Firm Performance, Strategic Management Journal, vol. 12, 1991, pp. 479-491.

Chakrabarti, A.K., The role of champion in product innovation, California Management Review, vol. 17(2), 1974, pp. 58-62.

Chakrabarti, A.K., Hauschildt, J., The division of labor in innovation management, R&D Management, vol. 19(2), 1989, pp. 161-171.

Chandler, A.D., Strategy and structure: Chapters in the history of the American enterprise, Cambridge, MA: MIT Press 1962.

Chesborough, H.W., Teece, D.J., When is virtual virtuous? Organizing for innovation, Harvard Business Review, January - February, 1996, pp. 65-73.

Christensen, C.M., The Innovator's Dilemma. When New Technologies Cause Great Firms to Fail, Boston, MA: Harvard Business School Press 1997.

Clark, K., Project scope and project performance: The effect of parts strategy and supplier involvement on product development, Management Science, vol. 35, 1989, pp. 1247-1263.

Clark, K., Fujimoto, T., Product development performance, Boston, MA: Harvard Business School Press 1991.

Cohen, W.M., Levinthal, D.A., Absorptive capacity: A new perspective on learning and innovation, Administrative Science Quarterly, vol. 35, 1990, pp. 128-152.

Coleman, J.J., The Semiconductor Industry Association and the Trade Dispute with Japan, Harvard Business School, Case, 9-387-191, 1987.

Conant, J.S., Mokwa, M.P., Varadarajan, P.R., Strategic Types, Distinctive Marketing Competencies and Organizational Performance: A Multiple Measures-Based Study, Strategic Management Journal, vol. 11, 1990, pp. 365-383.

Cooper, R.G., Kleinchmidt, E.J., New product success factors: A comparison of "kills" versus successes and failures, R&D Management, vol. 20, 1990, pp. 47-63.

Crawford, C.M., The hidden costs of accelerated product development, Journal of Product Innovation Management, vol. 9, 1992, pp. 188-199.

Cummings, T.G., Srivastva, S., Management of work: A socio-technical systems approach, Kent, OH: Kent State University Press 1977.

Daft, R.L., A dual-core model of organizational innovation, Academy of Management Journal, vol. 21, 1978, pp. 193-210.

Daft, R.L., Bureaucratic versus non-bureaucratic structure and the process of innovation and change, in: Bacharach, S.B. (Ed.), Research in the sociology of organizations, vol. 1, Greenwich, CT: JAI Press 1982, pp. 129-166.

Daft, R.L., Organization theory and design, Minneapolis, MN: West Publishing Company 1992.

Daft, R.L., Becker, S.W., The innovative organization, New York: Elsevier 1978.

Damanpour, F., The adoption of technological, administrative, and ancillary innovations: Impact of organizational factors, Journal of Management, vol. 13, 1987, pp. 675-688.

Damanpour, F., Organizational Innovation: A meta-analysis of effects of determinants and moderators, Academy of Management Journal, vol. 34(3), 1991, pp. 555-590.

Damanpour, F., Organizational size and innovation, Organization Studies, vol. 13, 1992, pp. 375-402.

Damanpour, F., Organizational complexity and innovation: Developing and testing multiple contingency models, Management Science, vol. 42, 1996, pp. 693-716.

Damanpour, F., Chaganti, R., Configurations in Public Organizations: An Empirical Study of Strategy-Structure Alignments in Public Libraries, Advances in Strategic Management, vol. 6, 1990, pp. 227-246.

Damanpour, F., Evan, W.M., Organizational innovation and performance: The problem of organizational lag, Administrative Science Quarterly, vol. 29, 1984, pp. 392-409.

Damanpour, F., Gopalakrishnan, S., Theories of organizational structure and innovation adoption: the role of environmental change, Journal of Engineering and Technology Management, vol. 15, 1998a, pp. 1-24.

Damanpour, F., Gopalakrishnan, S., The dynamics of the adoption of product and process innovations in organizations, working paper, Rutgers Faculty of Management, Newark, NJ 1998b.

Damanpour, F., Szabat, K.A., Evan, W.M., The relationship between types of innovations and organizational performance, Journal of Management Studies, vol. 26, 1989, pp. 587-601.

D'Aveni, R.A., Hypercompetition, New York: Free Press 1994.

Day, D., Raising Radicals: Different Processes for Championing Innovative Corporate Ventures, Organizational Science, vol. 5(2), 1994, pp. 148-172.

Deschamps, J.P., Nayak, P.R., Competing through products: Lessons from the winners, Columbia Journal of World Business, vol. 27(2), 1992, pp. 38-54.

Design Management Institute, Polaroid Corporation: Camera Design and Development 1984, Case 9-993-023, Boston, MA 1993.

Deutsche Bundesbank, Monatsberichte November, vol. 42-49, 1990-97.

Dewar, R.D., Dutton, J.E., The adoption of radical and incremental innovations: An empirical analysis, Management Science, vol. 32, 1986, pp. 1422-1433.

Dickson, K., Smith, H.L., Smith, S.L., Bridge Over Troubled Waters? Problems and Opportunities in Interfirm Research Collaboration, Technology Analysis & Strategic Management, vol. 3, 1991, pp. 143-156.

Dillard, J.P., Burgoon, M., Situational influences on the selection of compliance-gaining messages: Two tests of the predictive utility of the Cody-McLaughlin typology, Communication Monographs, vol. 52, 1985, pp. 289-304.

Dobberstein, N., Technologiekooperationen zwischen kleinen und großen Unternehmen - Eine transaktionskostentheoretische Perspektive, PhD.-Diss. University of Kiel 1992.

Dougherty, D., Hardy, C., Sustained product innovation in large, mature organizations: Overcoming innovation-organization problems, Academy of Management Journal, vol. 39, 1996, pp. 1120-1153.

Downs, G.W., Mohr, L.B., Conceptual issues in the study of innovation, Administrative Science Quarterly, vol. 21, 1976, pp. 700-714.

Doz, Y.L., Technology Partnerships Between Larger and Smaller Firms: Some Critical Issues, in: Contractor, F., Lorange, P. (Eds.), Cooperative Strategies in International Business, 1988, pp. 317-338.

Drake, B.H., Moberg, D.J., Communication influence attempts in dyads: Linguistic sedatives and palliatives, Academy of Management Review, vol. 11(3), 1986, pp. 567-584.

Ebadi, Y.M., Utterback, J.M., The Effects of Communication on Technological Innovation, in: Management Science, vol. 30, 1984, pp. 572-585.

Eisele, J., Erfolgsfaktoren des Joint Venture-Management, Wiesbaden: Gabler 1995.

Eisenhardt, K.M., Bourgeois, L.J., III, Politics of strategic decision making in high-velocity environments: Toward a midrange theory, Academy of Management Journal, vol. 31(4), 1988, pp. 737-770.

Eisenhardt, K.M, Tabrizi, B., Accelerating adaptive processes: Product innovation in the global computer industry, Administrative Science Quarterly, vol. 40, 1995, pp. 84-110.

Emery, F.E., Trist, E.L., Socio-technical systems, in: Churchman, C.W., Verhulst, M. (Eds.), Management science: Models and techniques, Oxford: Pergamon 1960, pp. 83-70.

Ernst, H., Patenting strategies in the German mechanical engineering industry and their relationship to company performance, Technovation, vol. 15(4), 1995, pp. 225-240.

Ernst, H., Patentinformationen für die strategische Planung von Forschung und Entwicklung, Wiesbaden: DUV 1996.

Ernst, H., The use of patent data for technological forecasting: The diffusion of CNC-Technology in the machine tool industry, Small Business Economics, vol. 9(4), 1997, pp. 361-381.

Ernst, H., Patent portfolios for strategic R&D planning, Journal of Engineering and Technology Management, 1998, forthcoming.

Ernst, H., Teichert, T., The R&D/Marketing interface and single informant bias in NPD research: An illustration of a benchmarking case study, Technovation, 1998, forthcoming.

Ettlie, J.E., A note on the relationship between managerial change values, innovative intentions and innovative technology outcomes in food sector firms, R&D-Management, vol. 13, 1983, pp. 231-244.

Ettlie, J.E., Taking charge of manufacturing. San Francisco, CA: Jossey-Bass 1988.

Ettlie, J.E., Product-process development integration in manufacturing, Management Science, vol. 41, 1995, pp. 1224-1237.

Ettlie, J.W., Bridge, W.P., O'Keefe, R.D., Organization strategy and structural differences for radical versus incremental innovation, Management Science, vol. 30(6), 1984, pp. 682-695.

Ettlie, J.E., Reza, E.M., Organizational integration and process innovation, Academy of Management Journal, vol. 35, 1992, pp. 795-827.

Evan, W.M., Organizational lag, Human Organizations, vol. 25, 1966, pp. 51-53.

Evan, W.M., Organization theory and organizational effectiveness: An exploratory analysis, Organization and Administrative Science, vol. 7, 1976, pp. 15-28.

Evan, W., Olk, P., R&D Consortia: A New U.S. Organizational Form, Sloan Management Review, vol. 37, 1990, pp. 37-46.

Eversheim, W., Technische Problemlösung, in: Eversheim, W., Schuh, G. (Eds.), Produktion und Management "Betriebshütte", part 1, 7. ed., Berlin et al.: Springer 1996, pp. 7-20.

Exhibition catalog of the Altona Museum, Hamburg, on: Frühe Formen der Werbung, 1996, catalog numbers 11.36 to 11.38.

Falbo, T., Peplau, L.A., Power strategies in intimate relations, Journal of Personality and Social Psychology, vol. 38, 1980, pp. 618-628.

Farr, C.M., Fischer, W.A., Managing International High Technology Cooperative Projects, R&D Management, vol. 22, 1992, pp. 55-67.

Faust, K., Neue technologische Trends im Licht der internationalen Patentstatistik und der Orientierung der deutschen Forschung, München: Ifo-Institut für Wirtschaftsforschung 1989.

FAZ Frankfurter Allgemeine Zeitung, Gardena Aktien zwischen 34 und 38 DM, Sept. 18, 1996.

FAZ Frankfurter Allgemeine Zeitung, hap., Technologiekonzern TRW baut Forschungszentrum in Deutschland, July 19, 1997.

FAZ Frankfurter Allgemeine Zeitung, Henkel klebt seit 75 Jahren, Jan. 9, 1998.

Fenneteau, H., Guibert, N., Trust in Buyer-Seller Relationships: Towards a Dynamic Classification of the Antecedents, in: Mazet, F., Salle, R., Valla, J.-P. (Eds), Interactions, Relationships and Networks in Business Markets, Proceedings of the 13th IMP Conference, Competitive Papers, Lyon 1997, pp. 217-248.

Fischer, W.A., Hamilton, W., McLaughlin, C., Zmud, R.W., The elusive product champion, Research Management, May - June, 1986, pp. 13-16.

Fisher, L.M., Chiron and Ciba-Geigy in collaboration with N.Y.U., New York Times (Late New York Edition), May 9, 1995, p. D8.

Fontanari, M., Kooperationsgestaltungsprozesse in Theorie und Praxis, Berlin: Duncker & Humblot 1996.

Fornell, C., Lorange, P., Roos, J., The Cooperative Venture Formation Process: A Latent Variable Structural Modeling Approach, Management Science, vol. 36, 1990, pp. 1246-1255.

Franks, I., Mayer, C., Ownership and Control, in: Siebert, H. (Ed.), Trends in Business Organisation, Tübingen: Mohr 1995, pp 171-200.

French, J.R.P., Raven, B., The bases of social power, in: Cartwright, D. (Ed.), Studies in Social Power, Ann Arbor: Institute for Social Research, University of Michigan 1959, pp.150-167.

Frost, P.J., Egri, C.P., Influence of political action on innovation, Part I, Leadership and Organization Development Journal, vol. 11(1), 1990, pp. 17-25.

Frost, P.J., Egri, C.P., The political process of innovation, in: Cummings, L.L., Staw, B.M. (Eds.), Research in organizational behavior, vol. 13, 1991, pp. 229-295.

Fujimoto, T., Comparing Performance and Organization of Product Development Across Firms, in: Eto, H. (Ed.), R&D Strategies in Japan, Amsterdam: North Holland 1993, pp. 143-175.

Gaitanides, M., Integrierte Belieferung - Eine ressourcenorientierte Erklärung der Entstehung von Systemlieferanten in der Automobilzulieferindustrie, Zeitschrift für Betriebswirtschaft, vol. 67., 1997, pp. 717-757.

Galbraith, J.R., Nathanson, D.A., Strategy implementation: The role of structure and processes, St. Paul, MN: West Publishing 1978.

Gannon, M.J., Smith, K.G., Grimm, C., An Organizational Information-Processing Profile of First Movers, Journal of Business Research, vol. 25, 1992, pp. 231-241.

Gemünden, H.G., Innovationsmarketing - Interaktionsbeziehungen zwischen Hersteller und Verwender innovativer Investitionsgüter, Tübingen: Mohr/Siebeck 1981.

Gemünden, H.G., Erfolgsfaktoren des Projektmanagements - eine kritische Bestandsaufnahme der empirischen Untersuchungen, Projektmanagement, vol. 1, 1990, pp. 4-15.

Gemünden, H.G., Zielbildung, in: Corsten, H., Reiß, M. (Eds.), Handbuch Unternehmensführung. Konzepte - Instrumente - Schnittstellen, Wiesbaden: Gabler 1995, pp. 251-266.

Gemünden, H.G., Heydebreck, P., Technological Interweavement - A Key Success Factor for New Technology-Based Firms, in: Sydow, J., Windeler, A. (Eds.), Management interorganisationaler Beziehungen, Opladen: Westdeutscher Verlag 1994, pp. 194-211.

Gemünden, H.G., Heydebreck, P., Technological Interweavement: A Means to Achieve Higher Efficiency in Production Processes, in: Oakey, R.P., Mukhtar, S.M (Eds.), New Technology-Based Firms in the 1990s, vol. III. London: Paul Chapman 1997, pp. 140-150.

Gemünden, H.G., Heydebreck, P., Herden, R., Technological Interweavement: A Means of Achieving Innovation Success, R&D Management, vol. 22, 1992, pp. 359-376.

Gemünden, H.G., Högl, M., Teamarbeit in innovativen Projekten: Eine kritische Bestandsaufnahme der empirischen Forschung, Zeitschrift für Personalforschung, vol. 12(3), 1998, pp. 277-301.

Gemünden, H.G., Lechler, T., Schlüsselfaktoren des Projekterfolges - Eine Bestandsaufnahme der empirischen Forschungsergebnisse, in: Knauth, P., Wollert, A. (Eds.), Praxishandbuch "Human Ressource Management" - Neue Formen betrieblicher Arbeitsorganisation und Mitarbeiterführung, Köln: Deutscher Wirtschaftsdienst, 1997, chapt. 7.7.2.2, pp. 1-30.

Gemünden, H.G., Ritter, T., Managing Technological Networks: The Concept of Network Competence, in: Gemünden, H.G., Ritter, T., Walter, A. (Eds.), Relationships and Networks in International Markets, Devon, UK: Elsevier Science 1998a, pp. 294-304.

Gemünden, H.G., Ritter, T., The Impact of Radical Environmental Change on a Company's Network Activities. An Empirical Study in East and West Germany, in: Urban, S. (Ed.), From Alliance Practices to Alliance Capitalism, Wiesbaden: Gabler 1998b, pp. 95-130.

Gemünden, H.G., Ritter, T., Heydebreck, P., Network Configuration and Innovation Success - An Empirical Analysis in German High-Tech Industries, International Journal of Research in Marketing, vol. 13, 1996, pp. 449-462.

Gemünden, H.G., Walter, A., Der Beziehungspromotor - Schlüsselperson für inter-organisationale Innovationsprozesse, in: Zeitschrift für Betriebswirtschaft, vol. 65, 1995, pp. 971 - 986.

Gemünden, H.G., Walter, A., Beziehungspromotoren - Schlüsselpersonen für zwischenbetriebliche Innovationsprozesse, in: Hauschildt, J., Gemünden, H.G. (Eds.), Promotoren - Champions der Innovation, Wiesbaden: Gabler 1998.

Germain, R., The role of context and structure in radical and incremental logistics innovation adoption, Journal of Business Research, vol. 35, 1996, pp. 117-127.

Gilbert, J.T., Profiting from innovation: Inventors and adopters, Industrial Management, vol. 37(4), 1995, pp. 28-32.

Ginsberg, A., Measuring and Modelling Changes in Strategy: Theoretical Foundations and Empirical Directions, Strategic Management Journal, vol. 9, 1988, pp. 559-575.

Glismann, H.H., Horn, E.-J., Comparative invention performance of major industrial countries: Patterns and explanations, Management Science, vol. 34(10), 1988, pp. 1169-1187.

Globe, S., Levy, G.W., Schwartz, C.M., Key Factors and Events in the Innovation Process, Research Management, vol. XVI, 1973, pp. 8-15.

Gopalakrishnan, S., Bierly, P., Organizational innovation and strategic choices: A knowledge based view, Academy of Management Best Paper Proceedings, 1997, pp. 242-246.

Gopalakrishnan, S., Bierly, P.E., Kessler, E.H., A re-examination of product and process innovation using a knowledge-based view, Journal of High Technology Management Research, to appear 1998.

Gopalakrishnan, S., Damanpour, F., Patterns of generation and adoption of innovations in organizations: Contingency models of innovation attributes, Journal of Engineering and Technology Management, vol. 11, 1994, pp. 95-116.

Goss, K.F., Consequences of diffusion of innovations, Rural Sociology, vol. 44, 1979, pp. 754-772.

GPO, Annual report (1994), München 1994.

Graen, G.B., Role making processes within complex organizations, in: Dunnette, M.D. (Ed.), Handbook of Industrial and Organizational Psychology, Chicago: Rand McNally 1976.

Graen, G.B., Cashman, J.F., A role making model of leadership in formal organizations: A developmental approach, in: Hunt J.G, Larson, L.L. (Eds.), Leadership Frontiers, Kent, OH: Comparative Administration Research Institute, Kent State University 1975.

Graen, G.B., Scandura, T., Toward a psychology of dyadic organizing, Research in Organizational Behavior, vol. 9, 1987, pp. 175-208.

Grant, R.M., The Resource-Based Theory of Competitive Advantage: Implications for Strategy Formulation, California Management Review, 1991, pp. 114-135.

Grant, R.M., Prospering in dynamically-competitive environments: Organizational capability as knowledge integration, Organization Science, vol. 7(4), 1996, pp. 375-387.

Grant, R.M., Contemporary Strategy Analysis, 3rd ed., Malden, MA: Blackwell Publishers 1998.

Graves, S.B., Why costs increase when projects accelerate, Research Technology Management, vol. 32(2), 1989, pp. 16-18.

Green, S.G., Welsh, A., Dehler, G.E., A Prospective Study of Industrial Innovation Medical Research Division, Unpublished manuscript, University of Cincinnati, College of Business Administration, Cincinnati, OH 1985.

Greve, H.R., Performance, Aspirations, and Risky Organizational Change, Administrative Science Quarterly, vol. 43, 1998, pp. 58-86.

Griliches, Z., Patent statistics as economic indicators: A survey, Journal of Economic Literature, vol. 18(4), 1990, pp. 1661-1707.

Griliches, Z., Pakes, A., Hall, B.H., The value of patents as indicators of inventive activity, Discussion Paper No. 1285, Cambridge, MA: Harvard Institute of Economic Research 1986.

Griset, P., Innovation and Radio Industry in Europe during the Interwar Period, in: Caron, F., Erker, P., Fischer, W. (Eds.), Innovations in the European Economy between the Wars, Berlin, New York: de Gruyter 1995, pp. 37-64.

Grossman, J.B., The supreme court and social change: A preliminary inquiry, American Behavioral Scientist, vol. 13, 1970, pp. 535-551.

Gutenberg, E., Der Einfluss der Betriebsgrösse auf die Kostengestaltung in Fertigungsbetrieben, Schweizerische Zeitschrift für Kaufmännisches Bildungswesen, vol. 50, 1956, pp. 1-10, 28-37.

Hage, J., Theories of organizations, New York: Wiley 1980.

Hagedoorn, J., Understanding the Rationale of Strategic Technology Partnering: Interorganizational Modes of Cooperation and Sectoral Differences, Strategic Management Journal, vol. 14, 1993, pp. 371-385.

Hagedoorn, J., Schakenraad, J., The Effects of Strategic Technology Alliances on Company Performance, Strategic Management Journal, vol. 15, 1994, pp. 247-266.

Håkansson, H., Industrial Technological Development: A Network Approach, London: Croom Helm 1987.

Håkansson, H., Corporate Technological Behaviour: Co-operation and Networks, London/New York: Routledge 1989.

Håkansson, H., Managing Cooperative Research and Development: Partner Selection and Contract Design, R&D Management, vol. 23, 1993, pp. 224-237.

Hall, B.H., Griliches, Z., Hausmann, J.A., Patents and R&D: Is there a lag?, International Economic Review, vol. 27(2), 1986, pp. 265-283.

Hambrick, D.C., Some Tests of the Effectiveness and Functional Attributes of Miles and Snow's Strategic Types, Academy of Management Journal, vol. 26, 1983, pp. 5-26.

Hambrick, D.C., MacMillan, I.C., Barbosa, R.R., Business unit strategy and changes in product R&D budgets, Management Science, vol. 29, 1983, pp. 757-769.

Hamel, G., Competition for competence and interpartner learning within international strategic alliances, Strategic Management Journal, vol. 12, 1991, pp. 83-103.

Hamel, G., Prahalad, C.K., Competing for the Future, Boston, MA: Harvard Business School Press 1994.

Handfield, R.B., Effects of concurrent engineering on make-to-order products, IEEE Transactions on Engineering Management, vol. 41, 1994, pp. 384-393.

Harhoff, D., Innovationsanreize in einem strukturellen Oligopolmodell, Zeitschrift für Wirtschafts- und Sozialwissenschaften, vol. 117(3), 1997, pp. 323-355.

Harhoff, D., Narin, F., Scherer, F.M., Vopel, K., Citation frequency and the value of patented innovation, WZB-Arbeitspapier FS IV, No. 26, Berlin 1997.

Harvey, M.G, Lusch, R.F., A Systematic Assessment of Potential International Strategic Alliance Partners, International Business Review, vol. 4, 1995, pp. 195-212.

Hauschildt, J., Entscheidungsziele. Zielbildung in innovativen Entscheidungsprozessen: theoretische Ansätze und empirische Prüfung. Tübingen: Mohr/ Siebeck 1977.

Hauschildt, J., Zur Messung des Innovationserfolges, Zeitschrift für Betriebswirtschaft, vol. 61, 1991, pp. 451-476.

Hauschildt, J., Erfolgs-, Finanz- und Bilanzanalyse, 3rd ed., Köln: O. Schmidt 1996.

Hauschildt, J., Innovationsmanagement, 2. ed., München: Vahlen 1997.

Hauschildt, J., Widerstand gegen Innovationen – konstruktiv oder destruktiv?, Zeitschrift für Betriebswirtschaft, to appear 1999.

Hauschildt, J., Chakrabarti, A.K., Arbeitsteilung im Innovationsmanagement - Forschungsergebnisse, Kriterien und Modelle, ZfO, vol. 57, 1988, pp. 378-388.

Hauschildt, J., Gemünden, H.G. (Eds.), Promotoren - Champions der Innovation, Wiesbaden: Gabler 1998.

Hauschildt, J., Grün, O. (Eds.), Ergebnisse Empirischer Betriebswirtschaftlicher Forschung: Zu einer Realtheorie der Unternehmung, Festschrift für E. Witte, Stuttgart: Schäffer-Poeschel 1993.

Hauschildt, J., Leker, J., Kreditwürdigkeitsprüfung, inkl. automatisierte, in: Handwörterbuch des Bank- und Finanzwesens, 2nd ed., Stuttgart: Schäffer-Poeschel 1995, col. 1321-1335.

Havelock, R.G., The Change Agent's Guide to Innovation, Englewood Cliffs, NJ: Educational Technology Publ. 1973.

Helfert, G., Teams im Relationship Marketing. Design effektiver Kundenbeziehungsteams, Wiesbaden: Gabler 1998.

Henderson, R., Clark, K.B., Architectural innovation: the reconfiguration of existing product technologies and the failure of established firms, Administrative Science Quaterly, vol. 35, 1990, pp. 9-30.

Henderson, R., Cockburn, I., Measuring competence? Exploring firm effects in pharmaceutical research, Strategic Management Journal, vol. 15 (Winter Special Issue), 1994, pp. 63-84.

Herbst, P.G., Socio-technical design, London: Tavistock 1974.

Heydebreck, P., Technologische Verflechtung: Ein Instrument zum Erreichen von Produkt- und Prozessinnovationserfolg. Frankfurt/Main: Lang 1996.

Hinkin, T.R., Schriesheim, C.A., Power and influence: The view from below, Personnel, vol. 65(5), 1988, pp. 47-50.

Hinkin, T.R., Schriesheim, C.A., Relationships between subordinate perceptions of supervisor influence tactics and attributed bases of supervisory power, Human Relations, vol. 43(3), 1990, pp. 221-237.

Hippel, E. von, Lead Users: Sources of Novel Product Concepts, Management Science, vol. 32, 1986, pp. 791-805.

Högl, M., Teamarbeit in innovativen Projekten. Einflußgrößen und Wirkungen, Wiesbaden: Gabler 1998.

Högl, M., Gemünden, H.G., Teamwork Quality and the Success of Innovations: An Empirical Investigation of Software Development Teams in Germany, Competitive Paper presented at the Annual Conference of the Academy of Management, San Diego, August, 1998a.

Högl, M., Gemünden, H.G., Team-Member Skills, Teamwork Quality and the Performance of Innovation Teams: An Empirical Investigation of Software Development Teams in Germany, Competitive Paper presented at the Product Development & Management Association (PDMA) 1998 International Research Conference, Atlanta, October, 1998b.

Högl, M., Gemünden, H.G., Determinanten und Wirkungen der Teamarbeit in innovativen Projekten: Eine theoretische und empirische Analyse der Teamarbeit, Zeitschrift für Betriebswirtschaft, to appear 1999.

Howell, J.M., Higgins, C.A., Champions of technological innovation, Administrative Science Quarterly, vol. 35(2), 1990, pp. 317-341.

Hull, F., Collins, P.D., Likers, J.K., Composite forms of organization as a strategy for concurrent engineering effectiveness, IEEE Transactions on Engineering Management, May, 1996.

Hull, F., Slowinsky, E., Partnering with Technology Entrepreneurs, Research Technology Management, vol. 33(6), 1990, pp. 16-20.

Iansiti, M., Shooting the rapids: Managing product development in turbulent environments, California Management Review, vol. 38(1), 1995, pp. 37-58.

Inkpen, A.C., Birkenshaw, J., International Joint Ventures and Performance: an Interorganizational Perspective, International Business Review, vol. 3, 1994, pp. 201-217.

Italiander, R., Ferdinand Graf von Zeppelin - Reitergeneral, Diplomat, Luftschiffpionier, Konstanz: Stadler 1986.

Izuchukwe, J., Architecture and process: The role of integrated systems in concurrent engineering introduction, Industrial Management, March/April, 1992, pp. 19-23.

Janis, J.L., Groupthink - Psychological Studies of Policy Decisions and Fiascoes, Boston et al. 1982 (1st. ed. 1972).

Javidan, M., Core Competence: What Does it Mean in Practice?, Long Range Planning, vol. 31, 1998, pp. 60-71.

Jensen, M.C., Takeovers: Their Causes and Consequences, Journal of Economic Perspectives, vol. 2, 1988, pp. 323-329.

Jensen, M.C., Ruback, R.S., The Market for Corporate Control: The Scientific Evidence, Journal of Financial Economics, vol. 10, 1983, pp. 5-50.

Kaiser, H.F., An Index of factor simplicity, Psychometrika, vol. 35, 1974, pp. 401-415.

Kaplan, S., Top Executives, Turnover and Firm Performance in Germany, Journal of Law, Economics and Organization, vol. 10, 1994, pp. 142-159.

Katz, R., Allen, T.J., Investigating the Not Invented Here (NIH) Syndrome: A Look at the Performance Tenure and Communication Patterns of 50 R&D Project Groups, R&D Management, vol. 12, 1982, pp. 7-19.

Keim, G., Projektleiter in der industriellen Forschung und Entwicklung - Theoretische Ansätze und empirische Prüfung, Wiesbaden: DUV 1997.

Kemp, R., Ghauri, N., Evaluating Performance in Inter-Firm Partnerships: The Case of International Joint Ventures, in: Mazet, F., Salle, R., Valla, J.-P. (Eds), Interactions, Relationships and Networks in Business Markets, Proceedings of the 13th IMP Conference, Competitive Papers, Lyon 1997, pp. 305-332.

Kesselring, F., Technische Kompositionslehre, Berlin et al.: Springer 1954.

Kessler, E.H., Chakrabarti, A.K., Innovation speed: A conceptual model of context, antecedents and outcomes, Academy of Management Review, vol. 21(4), 1996, pp. 1143-1191.

Kessler, E.H., Chakrabarti, A.K., Speeding up innovation: An empirical study of methods that increase the pace of new product innovations, Journal of Product Innovation Management, to appear 1998.

Keys, B., Case, T., How to Become an Influential Manager, Academy of Management Executive, vol. 4(4), 1990, pp. 38-51.

Kim, J., Mueller, C.W., Introduction to Factor Analysis, Beverly Hills, CA: Sage 1978.

Kimberly, J.R., Evanisko, M., Organizational innovation: The influence of individual, organizational, and contextual factors on hospital adoption of technological and administrative innovations, Academy of Management Journal, vol. 24, 1981, pp. 689-713.

Kipnis, D., The Power Holders, Chicago: University of Chicago Press 1976.

Kipnis, D., Castell, P.J., Gergen, M., Mauch, D., Metamorphic Effects of Power. Journal of Applied Psychology, vol. 61(2), 1979, pp. 127-135.

Kipnis, D., Schmidt, S.M., Profile of organizational influence strategies, San Diego: University Associates 1982.

Kipnis, D., Schmidt, S.M., Upward-influence styles: Relationship with performance evaluation, salary, and stress, Administrative Science Quarterly, vol. 33, 1988, pp. 528-542.

Kipnis, D., Schmidt, S.M., Price, K., Still, C., Why do I like three: Is it your performance or my orders? Journal of Applied Psychology, vol. 66, 1981, pp. 324-328.

Kipnis, D., Schmidt, S.M., Wilkinson, I., Intraorganizational influence tactics: Explorations in getting one's way, Journal of Applied Psychology, vol. 65, 1980, pp. 440-452.

Kirchmann, E.M.W., Innovationskooperation zwischen Herstellern und Anwendern, Wiesbaden: DUV 1994.

Kirchmann, E.M.W., Innovationskooperation zwischen Hersteller und Anwender, Zeitschrift für betriebswirtschaftliche Forschung, vol. 48, 1996, pp. 442-465.

Klöter, R., Opponenten im organisationalen Beschaffungsprozeß, Wiesbaden: Gabler 1997.

Knight, K.E., A descriptive model of intra-firm innovation process, Journal of Business, vol. 41, 1967, pp. 478-496.

Kogut, B., Zander, U., Knowledge of the firm, combinative capabilities, and the replication of technology, Organization Science, vol. 3(3), 1992, pp. 383-397.

Kölmel B., Erfolgsmessung europäischer kooperativer F&E-Projekte - eine empirische Analyse anhand von ESPRIT-Projekten, PhD.-Diss. University of Karlsruhe 1996.

König, H., Licht, G., Staat, M., F&E-Kooperationen und Innovationsaktivität, in: Gahlen, B., Ramser, H.J., Hesse, H. (Eds.), Ökonomische Probleme der europäischen Integration, Tübingen: Mohr/Siebeck 1993.

Kreiner, K., The Art of Managing EUREKA Projekts. Copenhagen: Copenhagen Business School 1994a.

Kreiner, K., EUREKA Projects and Contextual Turbulence - Managerial and Organizational Strategies for Survival and Achievement, Paper presented on the Oxford Conference on Management of Collaborative European Programs and Projects in Research and Training, April, 1994b.

Kropeit, G., Erfolgsfaktoren für die Gestaltung von FuE-Kooperationen, PhD.-Diss. TU Dresden 1998.

Kumar, N., Stern, L., Anderson, L.C., Conducting interorganizational research using key informants, Academy of Management Journal, vol. 36, 1993, pp. 1633-1651.

Küting, K., Weber, C.-P., Die Bilanzanalyse: Lehrbuch zur Beurteilung von Einzel- und Konzernabschlüssen, 3rd ed., Stuttgart: Schäffer-Poeschel 1997.

Kusunoki, K., Incapability of technological capability: A case study on product innovation in the Japanese facsimile industry, Journal of Product Innovation Management, vol. 4, 1997, pp. 368-382.

Lamb, D., Easton, S.M., Multiple Discovery. The pattern of scientific progress, Avebury Publ. Co., 1984.

Lambe, C.J., Spekman, R.E., Alliances, External Technology Acquisition, and Discontinuous Technological Change, Journal of Product Innovation Management, vol. 14, 1997, pp. 102-116.

Lange, V., Technologische Konkurrenzanalyse, Wiesbaden: DUV 1994.

Langrish, J., Gibbons, M., Evans, W.G., Jevons, F.R., Wealth from Knowledge - Studies of Innovation in Industry, London: Basingstroke 1972.

Lawless, M.W., Anderson, P.C., Generational technological change: Effects of innovation and local rivalry on performance, Academy of Management Journal, vol. 39, 1996, pp. 1185-1217.

Lawless, M.W., Price, L.L., An agency perspective on new technology champions, Organization Science, vol 3(3), 1992.

Lechler, T., Erfolgsfaktoren des Projektmanagements, Frankfurt am Main: Lang 1997.

Lechler, T., Gemünden, H.G., Kausalanalyse der Wirkungsstruktur der Erfolgsfaktoren des Projektmanagements - empirische Analyse eines konzeptionellen Bezugsrahmens, Die Betriebswirtschaft, vol. 58, 1998, pp. 435-450.

Lee, J., Small firms' innovation in two technological settings. Research Policy, vol. 24, 1995, pp. 391-401.

Lee, Michelle K., Lee, Mavis K., High Technology Consortia, IEEE Engineering Management Review, Spring, 1993, pp.77-88.

Lee U., Lee J., Bobe B., Technological Cooperation Between European and Korean Small Firms: the Patterns and Success Factors of Contracts, International Journal of Technology Management, vol. 8, 1993, pp.764- 781.

Lei, D., Hitt, M.A., Bettis, R., Dynamic core competencies through meta-learning and strategic context, Journal of Management, 1996.

Leiberman, M., Montgomery, D., First mover advantages, Strategic Management Journal, vol. 9, 1988, pp. 41-58.

Leins, H., Wissensbasierte Unternehmensanalyse - Effizienzsteigerung der Bonitätsbeurteilung im Firmenkundengeschäft, Wiesbaden: Gabler 1993.

Leker, J., Fraktionierende Frühdiagnose von Unternehmenskrisen: Bilanzanalysen in unterschiedlichen Krisenstadien, Köln: O. Schmidt 1993.

Lengnick-Hall, C.A., Innovation and competitive advantage: What we know and what we need to learn, Journal of Management, vol. 18, 1992, pp. 399-429.

Leonard-Barton, D., Wellsprings of Knowledge, Boston, MA: Harvard Business School Press 1995.

Liden, R.C., Mitchell, R.R., Ingratiatory behaviors in organizational settings, Academy of Management Review, vol. 13(4), 1988, pp. 572-587.

Lieberman, M., Montgomery, D., First-mover advantages, Strategic Management Journal, vol. 9, 1988, pp. 41-58.

Lindblom, C.E., The Science of Muddling through, Public Administration Review, vol. 19, 1959, pp. 79-88.

Linné, H., Magnaval, R., Removille, J., Key Factors for Industrial Partnership in EC Progammes, Luxemburg: Commission of the European Communities 1991.

Linné, H., Wahl geeigneter Kooperationspartner: Ein Beitrag zur strategischen Planung von F+E-Kooperationen. Frankfurt/Main: Lang 1993.

Loose, A., Sydow, J., Vertrauen und Ökonomie in Netzwerkbeziehungen - Strukturationstheoretische Betrachtungen, in: Sydow, J., Windeler, A. (Eds.), Management interorganisationaler Beziehungen, Opladen: Westdeutscher Verlag 1994, pp. 160-193.

Lubritz, S., Zur Bedeutung von Fits in Internationalen Strategischen Allianzen mittelständischer Unternehmen. Marketing- und Management-Transfer, Institut für Handel und internationales Marketing an der Universität des Saarlandes, Saarbrücken, vol. 10, 1996, pp.28-31.

Lynn, G.S., Mazzuca, M., Morone, J.G., Paulson, A.S., Learning Is the Critical Success Factor in Developing Truly New Products, Research Technology Management, vol. 41(3), 1998, pp. 45-51.

Lynn, G.S., Morone, J.G., Paulson, A.S., Marketing and discontinuous innovation: The probe and learn process, California Management Review, vol. 38, 1996, pp. 8-37.

Madhavan, R., Koka, B.R., Prescott, J.E., Networks in transition: How industry events (re)shape interfirm relationships, Strategic Management Journal, vol. 19(5), 1998, pp. 439-459.

Mahoney, J.T., The Resource-Based View within the Conservation of Strategic Management, Strategic Management Journal, vol. 13, 1992, pp. 363-380.

Maidique, M.A., Entrepreneurs, Champions, and Technological Innovation, Sloan Management Review, vol. 2, 1980, pp. 59-76.

Maidique, M.A., Patch, P., Corporate strategy and technology policy, in: Tushman, M.L., Moore, W.A. (Eds.), Readings in the Management of Innovation, Marchfield, MA: Pittman 1982.

Mairesse, J., Sassenou, M., R&D and productivity: A survey of econometric studies at the firm level, STI-Review, vol. 8, 1991, pp. 9-43.

Markham, S.K., Championing and antagonism as forms of political behavior: An R&D perspective, Organization Science (in press).

Markham, S.K., Green, S.G., Basu, R., Champions and antagonists: Relationships with R&D project characteristics and management, Journal of Engineering and Technology Management, vol. 8(3+4), 1991, pp. 217-242.

Literature 337

Markham, S.K., Griffin, A., The breakfast of champions: Associations between champions and product development environments, practices, and performance, Journal of Product Innovation Management, vol. 15(5), 1998, pp. 436-454.

Marmor, A.C., Lawson, W.S., Terapane, J.F., The technology assessment and forecast program of the United States Patent and Trademark Office, World Patent Information, vol. 1(1), 1979, pp. 15-23.

Mayer, R.C., Davis, J.H., Schoorman, F.D., An Integrative Model of Organizational Trust, Academy of Management Review, vol. 20, 1995, pp. 709-733.

McDonough, E.F., Faster new product development: Investigating the effects of technology and characteristics of the project leader and team, Journal of Product Innovation Management, vol. 10, 1993, pp. 241-250.

McDonough, E.F., Barczak, G., Speeding up new product development: The effects of leadership style and source of technology, Journal of Product Innovation Management, vol. 8, 1991, pp. 203-211.

McGahan, A.M., Industry structure and competitive advantage, Harvard Business Review, November - December, 1994, pp. 115-124.

Meyer, A.D., Adapting to environmental jolt, Administrative Science Quarterly, vol. 27, 1982, pp. 515-537.

Meyer, C., Fast cycle time: How to align purpose, strategy, and structure for speed, New York: Free Press 1993.

Meyer, M., Ökonomische Organisation der Industrie. Netzwerkarrangements zwischen Markt und Unternehmung, Wiesbaden: Gabler 1994.

Mezias, S.J., Glynn, M.A., The three faces of corporate renewal: Institution, evolution, and revolution, Strategic Management Journal, vol. 14, 1993, pp. 77-101.

Miles, R.E., Snow, C.C., Organizational strategy, structure and process, New York: McGraw-Hill 1978.

Miles, R.H., Macroorganizational behavior, Santa Monica, CA: Goodyear 1980.

Miller, D., Friesen, P.H., Momentum and Revolution in Organizational Adaption, Academy of Management Journal, vol. 23, 1980, pp. 591-614.

Miller, G.A., Contay, J., Differentiations in organizations: Replication and cumulation, Social Forces, vol. 59, 1980, pp. 265-274.

Miller, M.D., Friendship, power, and the language of compliance-gaining, Journal of Language and Social Psychology, vol. 1, 1982, pp. 111-121.

Miller, R., The new agenda for R&D: strategy and integration, International Journal of Technology Management, vol. 10, 1995, p. 511-524.

Millson, M.R., Raj, S.P., Wilemon, D., A survey of major approaches for accelerating new product development, Journal of Product Innovation Management, vol. 9, 1992, pp. 53-69.

Mintzberg, H., Strategy formation: Schools of thought, in: Fredrickson, J.F. (Ed.), Perspectives on Strategic Management, Philadelphia: Harper Business 1990.

Möhrle, M.G., Voigt, I., Das FuE-Programm-Portfolio in praktischer Erprobung, Zeitschrift für Betriebswirtschaft, vol. 63(10), 1993, pp. 973-992.

Mohr, J., Spekman, R., Characteristics of partnership success: Partnership attributes, communication behavior, and conflict resolution techniques, Strategic Management Journal, vol. 15(2), 1994, pp. 135-152.

Moorman, C., Zaltman, G., Deshpandé, R., Relationships Between Providers and Users of Market Research: The Dynamics of Trust Within and Between Organizations, Journal of Marketing Research, vol. 29, 1992, pp. 314-328.

Morbey, G.K., Reithner, R.M., How R&D affects sales growth, productivity and profitability, Research Technology Management, May - June, 1990, pp. 11-14.

Morgan, R., Hunt, S., The Commitment-Trust Theory of Relationship Marketing, Journal of Marketing, vol. 58, 1994, pp. 20-38.

Myers, S., Marquis, D.G., Successful industrial innovations, Washington, D.C.: National Science Foundation 1969.

Narin, F., Noma, E., Perry, R., Patents as indicators of corporate technological strength, Research Policy, vol. 16, 1987, pp. 143-155.

Neter, J., Wasserman, W., Kutner, M.H., Applied statistical linear models (3rd ed.), Homewood, Ill.: Irwin 1990.

Nijssen, E.J., Arbouw, A.R.L., Commandeur, H.R., Accelerating new product development: A preliminary empirical test of a hierarchy of implementation, Journal of Product Innovation Management, vol. 12, 1995, pp. 99-109.

Niosi, J., Bergeron, M., Technical Alliances in the Canadian Electronics Industry: An Empirical Analysis, Technovation, vol. 12, 1992, pp. 309-322.

Nolan, V., The Innovator's Handbook - The Skills of Innovative Management, Problem Solving, Communication and Teamwork, London: Sphere Books 1989.

Nonaka, I., A dynamic theory of knowledge creation, Organization Science, vol. 5(1), 1994, pp. 14-37.

Nord, W.R., Tucker, S., Implementing routine and radical innovation, Lexington, MA: Lexington Books 1987.

Norling, P.M., Statz, R.J., How Discontinuous Innovation Really Happens, Research Technology Management, vol. 41(3), 1998, pp. 41-44.

Normann, R., Organizational innovativeness: Product variation and reorientation, Administrative Science Quarterly, vol. 16, 1971, pp. 203-215.

Norusis, M.J., SPSS Inc., SPSS Professional Statistics 6.1, 1994.

Nueno, P., Oosterveld, J., Managing Technology Alliances, Long Range Planning, vol. 21(3), 1988, pp.11-17.

Office of Technology Assessment, U.S. Congress, Pharmaceutical R&D: Costs, Risks and Rewards, OTA-H-522, Washington, DC: US Government Printing Office 1993.

Ogburn, W.F., Social change, New York: Viking 1922.

Ohmae, K., The global logic of strategic alliances, Harvard Business Review, March - April, 1989, pp. 143-154.

Olk, P., Young, C., Why Members Stay in or Leave an R&D Consortium: Performance and Conditions of Membership as Determinants of Continuity, Strategic Management Journal, vol. 18, 1997, pp. 855-877.

Ormala, G., Evaluation of EUREKA Industrial and Economics Effects, Brussels 1993.

Paganetto, L. et al., The System Company, draft, in: Danielmeyer, H.G. (Ed.), Company of the Future, Berlin et al.: Springer, to appear.

Page, A., Assessing new product development practices and performance: Establishing crucial norms, Journal of Product Innovation Management, vol. 10, 1993, pp. 273-290.

Pakes, A., Griliches, Z., Patents and R&D at the firm level: A first look, in: Griliches, Z. (Ed.), R&D, patents and productivity, Chicago: The University of Chicago Press 1984, pp. 55-72.

Parnell, J.A., Consistency versus Flexibility: Does Strategic Change Really Improve Performance?, American Business Review, vol. 12, 1994, pp. 22-30.

Parnell, J.A., Performance´s Influence on Strategic Change: A Longitudinal Assessment, Scandinavian Journal of Management, vol. 14, 1998, pp. 19-36.

Pavitt, K., Uses and abuses of patent statistics, in: van Raan, A.F.J. (Ed.), Handbook of Quantitative Studies of Science and Technolgy, Amsterdam: Elsevier 1988, pp. 509-536.

Pavitt, K., Basic Sciences and Innovation, UNESCO World Science Report 1993, London 1993.

Pelz, D.C., Andrews, F.M., Scientists in Organizations: Productive Climates for Research and Development (revised edition), Ann Arbor: Institute for Social Research, University of Michigan 1976.

Penrose, E.T., The Theory of the Growth of the Firm, Oxford: Basil Blackwell 1959.

Penzias, A., Ideas and Information. Managing in a High-Tech World, New York, London: Norton 1989.

Perry, T.S., Managed chaos allows more creativity, Research Technology Management, vol. 38, 1995, pp. 14-17.

Pfeiffer, W., Schäffner, G.J., Schneider, W., Amler, R., Technologie-Portfolio zum Management strategischer Zukunftsgeschäftsfelder, 6th ed., Göttingen: Vandenhoek & Ruprecht 1991.

Pfeiffer, W., Schneider, W., Dögl, R., Technologie-Portfolio-Management, in: Staudt, E. (Ed.), Das Management von Innovationen, Frankfurt: Frankfurter Allgemeine Zeitung 1986, pp. 107-124.

Pfeiffer, W., Weiß, E., Methoden zur Analyse und Bewertung technologischer Alternativen, in: Zahn, E. (Ed.), Handbuch Technologiemanagement, Stuttgart: Schäffer-Poeschel 1995, pp. 663-679.

Picot, A., Transaktionskostenansatz, Handwörterbuch der Betriebswirtschaft, 5th ed., 3rd vol., Stuttgart: Schäffer-Poeschel 1993, col. 4194-4204.

Pisano, G.P., Using equity participation to support exchange: Evidence from the biotechnology industry, Journal of Law, Economics and Organization, vol. 5 (1), 1989, pp. 109-125.

Pisano, G.P., The governance of innovation: Vertical integration and collaborative arrangements in the biotechnology industry, Research Policy, vol. 20, 1991, pp. 237-249.

Pisano, G.P., Knowledge, integration, and the locus of learning: An empirical analysis of process development, Strategic Management Journal, vol. 15 (Winter Special Issue), 1994, pp.85-100.

Pisano, G.P., Wheelwright, S.C., The new logic of high-tech R&D, Harvard Business Review, September - October, 1995, pp. 93-105.

Pitts, C.E., For project managers: An inquiry into the delicate art and science of influencing others, Project Management Journal, vol. 21(1), 1990, pp. 21-42.

Plötner, O., Das Vertrauen des Kunden. Relevanz, Aufbau und Steuerung auf industriellen Märkten, Wiesbaden: Gabler 1995.

Poensgen, O.H., Fluktuation, Amtszeit und weitere Karriere von Vorstandsmitgliedern, Die Betriebswirtschaft, vol. 42, 1982, pp. 177-195.

Port, O., Carey, J., Getting to 'Eureka', Business Week, vol. 10, Nov., 1997, pp. 72-75.

Porter, M.E., Competitive Strategy, New York: Free Press 1980.

Powell, W.W., Learning from collaboration: Knowledge and networks in the biotechnology and pharmaceutical industries, California Management Review, vol. 40(3), 1998, pp. 228-240.

Powell, W.W., Brantley, P., Competitive cooperation in biotechnology: Learning through networks, in: Nohria, N., Eccles, R.G. (Eds.), Networks and organizations, Boston: Harvard Business School Press 1992, pp. 366-394.

Powell, W.W., Koput, K.W., Smith-Doerr, L., Interorganizational collaboration and the locus of innovation: Networks of learning in biotechnology, Administrative Science Quarterly, vol. 41, 1996, pp. 116-145.

Prahalad, C.K., Hamel, G., The Core Competence of the Corporation, Harvard Business Review, 1990, pp. 79-91.

Quinn, J.B., Strategic change: Logical incrementalism, Homewood: Irwin 1980.

Raffée, H., Eisele, J., Joint Ventures - Nur die Hälfte floriert, Harvard Business Manager, vol. 3, 1994, pp.17-21.

Raven, B.H., The comparative analysis of power and influence, in: Tedeschi, J.T. (Ed.), Perspectives in Social Power, Chicago: Aldine 1974.

Reed, R., Lemak, D.J., Montgomery, J.C., Beyond process: TQM content and firm performance, Academy of Management Review, vol. 21(1), 1996, pp. 173-202.

Reiche, D., Selzer, P., Berührungsangst, Manager Magazin, vol. 11, 1995, pp. 270-281.

Rice, M.P., O'Connor, G.C., Peters, L.S., Morone, J.G., Managing Discontinuous Innovation, Research Technology Management, vol. 41(3), 1998, pp. 52-58.

Ring, P.S., van de Ven, A., Development Processes of Cooperative Interorganizational Relationships, Academy of Management Review, vol. 19, 1994, pp. 90-118.

Ritter, T., Innovations-Erfolg durch Netzwerk-Kompetenz. Eine theoretische und empirische Analyse zum effektiven Management von Unternehmensnetzwerken, PhD.-Diss. Karlsruhe 1998.

Roberts, E.B., What we've learned: Managing invention and innovation, Research Technology Management, Jan.-Feb., 1988.

Roberts, E.B., Entrepreneurs in high technology: Lessons from MIT and beyond, New York: Oxford University Press 1991.

Roberts, E.B., Fusfeld, A.R., Critical function: Needed roles in the innovation process, in: Katz, R. (Ed.), Career issues in human resource management, Englewood Cliffs, NJ: Prentice-Hall, Inc. 1981, pp. 182-207.

Roberts, E.B., Fusfeld, A.R., Staffing the Innovative Technology-Based Organization, Sloan Management Review, vol. 22, 1981, pp. 19-34.

Robinson, E.A.G., Betriebsgröße und Produktionskosten, Wien: Springer 1936.

Rogers, E.M., Diffusion of innovations, 4th ed., New York: Free Press 1995.

Rogers, E.M., Shoemaker, F.F., Communication of Innovations. A Cross-Cultural Approach, New York: Free Press 1971.

Roos, J., Cooperative Venture Formation Processes: Characteristics and Impact on Performance, PhD.-Diss. Institute of International Business, Stockholm School of Economics 1989.

Rosenau, M.D., From experience: Faster new product development, Journal of Product Innovation Management, vol. 5., 1988, pp. 150-153.

Rosenberg, N., Inside the black box: Technology and economics, London: Cambridge University Press 1982.

Rosenthal, S.R., Effective product design and development: How to cut lead time and increase customer satisfaction, Homewood, Ill.: Business One Irwin 1992.

Rotering, C., Forschungs- und Entwicklungskooperationen zwischen Unternehmen - Eine empirische Analyse, Stuttgart: Poeschel 1990.

Rothwell, R., Freeman, C., Horsley, A., Jervis, V., Robertson, A.B., Townsend, J., SAPPHO updated-project SAPPHO phase II, Research Policy, vol. 3, 1974, pp. 258-291.

Rothwell, R., Robertson, A.B., The role of communications in technological innovations. Research Policy, vol. 2, 1973, pp. 204-225.

Rousseau, D.M., Sim, B., Sitkin, R.S., Camerer, C., Not So Different After All: A Cross-cultural View of Trust, Academy of Management Journal, vol. 23, 1998, pp. 393-404.

Rubenstein, A.H., Managing Technology in the Decentralized Firm, New York et al.: Wiley ETM 1989.

Rubenstein, A.H., Chakrabarti, A.K., O'Keefe, R.D., Souder, W.E., Young, H.C., Factors influencing innovation success at the project level, Research Management, vol. 19(3), 1976, pp. 15-20.

Rycroft, R.W., Kash, D.E., Complex technology and community: Implications for policy and social change, Research Policy, vol. 23, 1994, pp. 613-626.

Saad, A., Anbahnung und Erfolg von europäischen kooperativen F&E-Projekten, Frankfurt/Main: Lang 1998.

Sakakibara, K., R&D Cooperation Among Competitors: A Case Study of the VLSI Semiconductor Research Project in Japan, Journal of Engineering and Technology Management, vol. 10, 1993, pp. 393-407.

Samuels, J.M., Brayshaw, R.E., Craner, J.M., Financial Statement Analysis in Europe, London: Chapman & Hall 1995.

Sandig, C., Gewinn und Sicherheit in der Betriebspolitik - Das Treiben und das Bremsen im Betriebe, in: Zeitschrift für Betriebswirtschaft, vol. 10, 1933, pp. 349-360.

Schankerman, M., Measurement of the value of patent rights and inventive output using patent renewal data, STI-Review, vol. 8, 1991, pp. 101-122.

Schankerman, M., Pakes, A., Estimates of the value of patent rights in European countries during the post-1950 period, The Economic Journal, vol. 96, 1986, pp. 1052-1076.

Scherer, F.M., The propensity to patent, International Journal of Industrial Organization, vol. 1, 1983, pp. 107-128.

Schilit, W.K., An examination of the influence of middle-level managers in formulating and implementing strategic decisions, Journal of Management Studies, vol. 24(3), 1987, pp. 271-293.

Schilit, W.K., Locke, E., A study of upward influence in organizations, Administrative Science Quarterly, vol. 27, 1982, pp. 304-316.

Schmeisser, W.: Erfinder und Innovation - Widerstände im Inventionsprozeß, unter besonderer Berücksichtigung der Stellung und Bedeutung von Erfindern im Innovationsprozeß, PhD.-Diss. Duisburg 1984.

Schmoch, U., Wettbewerbsvorsprung durch Patentinformationen. Handbuch für die Recherchepraxis, Köln: TÜV Rheinland 1990.

Schmoch, U., Grupp, H., Mannsbart, W., Schwitalla, B., Technikprognosen mit Patentindikatoren, Köln: TÜV Rheinland 1988.

Schoeffler, S., Buzzell, R.D., Heany, D.F., Impact of strategic planning on profit performance, Harvard Business Review, vol. 52, 1974, pp. 137-145.

Schon, D.A., Champions for Radical New Inventions, Harvard Business Review, vol. 41, 1963, pp. 77-86.

Schrader, S., Spitzenführungskräfte, Unternehmensstrategie und Unternehmenserfolg, Tübingen: Mohr/Siebeck 1995a.

Schrader, S., Spitzenführungskräfte: Sind Techniker oder Wissenschaftler gefragt?, Technologie & Management, vol. 44(1), 1995b, pp. 9-14.

Schrader, S., Lüthje, C., Das Ausscheiden der Spitzenführungskraft aus der Unternehmung: Eine empirische Analyse, Zeitschrift für Betriebswirtschaft, vol. 65, 1993, pp. 467-493.

Schroeder, D.M., A dynamic perspective on the impact of process innovation upon competitive strategies, Strategic Management Journal, vol. 11, 1990, pp. 25-41.

Schumpeter, J., Theorie der wirtschaftlichen Entwicklung, Leipzig: Duncker & Humblot 1912.

Schumpeter, J., Theorie der wirtschaftlichen Entwicklung - Eine Untersuchung über Unternehmergewinn, Kapital, Kredit, Zins und den Konjunkturzyklus, 3. ed., Leipzig: Duncker & Humblot 1931.

Schwenk, C., Effects of devil's advocacy and dialectical inquiry on decision making: A meta-analysis, in: Organizational Behavior and Human Decision Processes, vol. 47, 1990, pp. 161-176.

Scott, S., Social Identification Effects in Product and Process Development Teams, Journal of Engineering and Technology Management, vol. 14, 1997, pp. 97-127.

Seers, A., Team-member exchange quality: A new construct for role-making research, Organizational Behavior and Human Decision Processes, vol. 43, 1989, pp. 118-135.

Sen, F., Chakrabarti, A.K., Acquisition and implementation of external technology: a framework for greater effectiveness, in: Proceedings Northeastern Meeting of Decision Sciences Institute, Williamsburg, VA, March 1986.

Shane, S.A., Venkataraman, S., MacMillan, I.C., The effects of cultural differences on new technology championing behavior within firms, The Journal of High Technology Management Research, Fall 1994.

Shina, D.K., Cusumano, M.A., Complementary Resources and Cooperative Research: A Model of Research Joint Ventures Among Competitors, Management Science, vol. 37, 1991, pp. 1091-1106.

Singh, K., The concept and implications of technological complexity for organisations, Proceedings of the Academy of Management Meeting, Michigan: School of Business, here quoted from: Tidd, J., Complexity, Networks & Learning: Integrative Themes for Research on Innovation Management, International Journal of Innovation Management, vol. 1, 1997, pp. 1-21.

Smith, J.B., Barclay, D.W., The Effects of Organizational Differences and Trust on the Effectiveness of Selling Partner Relationships, Journal of Marketing, vol. 61, 1997, pp. 3-21.

Smith, J.J., McKeon, J.E., Hoy, K.L., Boysen, R.L., Shechter, L., Roberts, E.B., Lesson From 10 Case Studies in Innovation, Research Management, September - October, 1984, pp. 23-27.

Smith, P.G., Reinertsen, D.G., Developing products in half the time, New York: Van Nostrand Reinhold 1991.

Snow, C.C., Hambrick, D.C., Measuring Organizational Strategies: Some Theoretical and Methodological Problems, Academy of Management Review, vol. 5, 1980, pp. 527-538.

Söllner, A., Commitment in Geschäftsbeziehungen: das Beispiel Lean Production, Wiesbaden: Gabler 1993.

Soete, L., The impact of technological innovation on international trade patterns: The evidence reconsidered, Research Policy, vol. 16, 1987, pp. 101-130.

Solow, R.M., Technical change and the aggregate production function, The Review of Economics and Statistics, vol. 39, 1957, pp. 312-320.

Sommerlatte, T., Management von Forschung und Entwicklung, in: Zahn, E. (Ed.), Handbuch Technologiemanagement, Stuttgart: Schäffer-Poeschel 1995, pp. 323-334.

Souder, W.E., Chakrabarti, A.K., Managing the coordination of marketing and R&D in the innovation process, TIMS Studies in the Management Sciences, vol. 15, 1980, pp. 135-150.

Souder, W.E., Nassar, S., Managing R&D Consortia for Success, Research Technology Management, September - October, 1990, pp. 44-50.

Specht, G., Beckmann, C., F&E-Management, Stuttgart: Schaeffer-Poeschel 1996.

Spender, J.-C., Industry Recipes: An Inquiry into the Nature and Sources of Managerial Judgement, New York: Blackwell 1989.

Spero, D.M., Patent Protection or Piracy - A CEO Views Japan, Harvard Business Review, vol. 68(5), 1990, pp. 58-67.

Staudt, E., Innovationsbarrieren und ihre Überwindung - Thesen aus einzelwirtschaftlicher Sicht, in: Giersch, H. (Ed.), Probleme und Perspektiven der weltwirtschaftlichen Entwicklung, Schriften des Vereins für Sozialpolitik, N. F., vol. 148, Berlin/München: Duncker & Humblot 1985, pp. 349-366.

Steinle, C.: Konfliktmanagement, in: Wittmann, W. et al. (Eds.), Handwörterbuch der Betriebswirtschaft, 5. ed., Stuttgart: Schäffer-Poeschel 1993, col. 2200-2216.

Strutton, H.D., Lumpkin, J.R., Vitell, S.J., An applied investigation of Rogers and Shoemaker's perceived innovation attribute typology when marketing to elderly consumers, Journal of Applied Business Research, vol. 10, 1994, pp. 118-131.

Sullivan, J.J., Albrecht, T.L., Taylor, S, Process, Organizational, Relational, and Personal Determinants of Managerial Compliance-gaining Communication Strategies, The Journal of Business Communication, vol. 27(4), 1990, pp. 333-355.

Swanger, C.C., Maidique, M.A., Apple Computer: The First Ten Years, Case S-BP-245, Stanford, CA 1985.

Swanson, E.B., Information systems innovation among organizations, Management Science, vol. 40, 1994, pp. 1069-1092.

Takeuchi, H., Nonaka, I., The new product development game, Harvard Business Review, vol. 64(1), 1986, pp. 137-146.

Teece, D.J., Profiting from technological innovation: Implications for integration, collaboration, licensing, and public policy, in: Teece, D.J. (Ed.), The Competitive Challenge, Cambridge, MA: Ballinger 1987, pp. 185-219.

Teichert, T. Erfolgspotential internationaler F&E-Kooperationen, Wiesbaden: DUV 1994.

Tidd, J., Development of novel products through interorganizational and intra-organizational networks: The case of home automation, Journal of Product Innovation Management, vol. 12, 1995, pp. 307-322.

Tidd, J., Bessant, J., Pavitt, K., Managing Innovation, Integrating Technological, Market and Organizational Change, Chichester et al.: Wiley 1997.

Tucci, C.L., Firm Heterogeneity and Performance of International Strategic Technology Alliances, Strategic Management Society Annual Meeting in Mexico City, Draft 1996.

Tucker, L.R. Koopman, R.F., Linn, R.L., Evaluation of factor analytic research procedures by means of simulated correlation matrices, Psychometrika, vol. 34, 1969, pp. 421-459.

Turner, R.H., Role taking: Process versus conformity, in: Rose, A.M. (Ed.), Human behavior and social processes: An interactionist approach, Boston: Houghton Mifflin 1962.

Tushman, M.L., Anderson, P., Technological discontinuities and organizational environments, Administrative Science Quarterly, vol. 31, 1986, pp. 439-465.

Tushman, M.L., Anderson, P., Managing strategic innovation and change, New York: Oxford 1997.

Tushman, M.L., O'Reilly, C.A., Ambidextrous organization: Managing evolutionary and revolutionary change, California Management Review, vol. 38 (summer), 1996, pp. 8-30.

Tushman, M.L., O'Reilly, C.A., Winning through innovation, Boston, MA: Harvard Business School Press 1997.

Tushman, M.L., Romanelli, E., Organizational Evolution: A Metamorphosis Model of Convergence and Reorientation, Research in Organizational Behavior, vol. 7, 1985, pp. 171-222.

Utterback, J.M., Mastering the Dynamics of Innovation: how companies can seize opportunities in the face of technological change, Boston, MA: Harvard Business School Press 1994.

Utterback, J.M., Abernathy, W.J., A dynamic model of process and product innovation, Omega, vol. 3, 1975, pp. 639-656.

Valacich, J.S., Schwenk, C., Structuring Conflict in Individual Face-to-Face, and Computer-Mediated Group Decision Making: Carping Versus Objective Devil's Advocacy, in: Decision Sciences, vol. 26, 1995, pp. 369-393.

Van de Ven, A., Central problem in the management of innovation, Management Science, vol. 32(5), 1986, pp. 590-607.

Van de Ven, A., Grazman, D.N., Technical Innovation, Learning, and Leadership, in: Garud, R., Nayyar, P., Shapira, Z. (Eds), Technological Oversights and Foresight, Cambridge: University Press 1995.

Vecchio, R.P., Sussmann, M., Choice of influence tactics: Individual and organizational determinants, Journal of Organizational Behavior, vol. 12(1), 1991, pp. 73-80.

Venkataraman, S., MacGrath, R.G., MacMillian, I.C., Progress in Research on Corporate Venturing, in: Sexton, D.L., Kasarda, J.D. (Eds.), The State of the Art of Entrepreneurship, Boston, MA: PWS Kent 1992.

Vesey, J.T., The new competitors: They think in terms of speed-to-market, Academy of Management Executive, vol. 5(2), 1991, pp. 23-33.

Walter, A., Der Beziehungspromotor. Ein personaler Gestaltungsansatz für erfolgreiches Relationship Marketing, Wiesbaden: Gabler 1998.

Walter, A., Gemünden, H.G., Beziehungspromotoren als Förderer inter-organisationaler Austauschprozesse, in: Hauschildt, J., Gemünden, H.G. (Eds.), Promotoren - Champions der Innovation, Wiesbaden: Gabler 1998, pp. 133-158.

Waudig, D., Verlauf und Erfolg kooperativer Innovationsprozesse zwischen Hochschule und Industrie, PhD.-Diss. University of Karlsruhe 1994.

Webb, J., Dawson, P., Measure for Measure: Strategic Change in an Electronic Instruments Corporation, Journal of Management Studies, vol. 28, 1991, pp. 191-206.

Wetzel, W., Zum Problem der unverzerrten Schätzung von Wachstumsraten, in: Schilcher, R. (Ed.), Wirtschaftswachstum. Beiträge zur ökonomischen Theorie und Politik, Berlin: de Gruyter 1964, pp. 131-137.

Wheelwright, S.C., Clark, K.B., Revolutionizing product development, New York: Free Press 1992.

Wimmer, W., Wir haben fast immer etwas Neues, Berlin: Duncker & Humblot 1994.

Wissema, J.G., Euser, L., Successful Innovation through Inter-Company Networks, Longe Range Planning, vol. 24, 1991, pp. 33-39.

Witte, E., Organisation für Innovationsentscheidungen: Das Promotoren-Modell, Göttingen: Schwartz 1973.

Witte, E., Kraft und Gegenkraft im Entscheidungsprozeß, in: ZfB, vol. 46, 1976, pp. 319-326.

Witte, E., Power and Innovation: a Two-Center-Theory, International Studies of Management and Organization, vol. 7, 1977, pp. 47-70.

Witte, E., Der Zusammenhang zwischen nachrichtentechnischen Innovationen und Veränderungen der Marktordnung, Sitzungsberichte der Bayerischen Akademie der Wissenschaften, vol. 5, 1997.

Wolfe, R.A., Organizational innovation: Review, critique, and suggested research directions, Journal of Management Studies, vol. 31, 1994, pp. 405-431.

Wolff, H., Becher, G., Delpho, H., Kuhlmann, S., Kuntze, U., Stock, J., F&E-Kooperation von kleinen und mittleren Unternehmen, Heidelberg: Physica 1994.

Wurche, S., Vertrauen und ökonomische Rationalität in kooperativen Interorganisationsbeziehungen, in: Sydow, J., Windeler, A. (Eds.), Management interorganisationaler Beziehungen, Opladen: Westdeutscher Verlag 1994, pp. 142-159.

Yoffie, D.B., Cohn, J., Levy, D., Apple Computer 1992, Harvard Business School, Case 9-792-081, 1992.

Yoffie, D.B., Pearson, A.E., The Transformation of IBM, Harvard Business School, Case 9-391-073, 1991.

Yukl, G., Falbe, C.M., Importance of Different Power Sources in Downward and Lateral Relations, Journal of Applied Psychology, vol. 76(3), 1991, pp. 416-423.

Zahra, S.A., Covin, J.G., The financial implications of fit between competitive strategy and innovation types and sources, Journal of High Technology Management Research, vol. 5, 1994, pp. 183-211.

Zairi, M., Youssef, M.A., Quality functional deployment, International Journal of Quality and Reliability Management, vol. 12(6), 1995, pp. 9-23.

Zaltman, G., Duncan, R., Holbeck, J., Innovations and organizations. New York: Wiley 1973, 1984.

Zhu, Z., Heady, R.B., A simplified method of evaluating PERT/CPM network parameters, IEEE Transactions on Engineering Management, vol. 41, 1994, pp. 426-430.

Zöfel, P., Statistik in der Praxis, 2nd ed., Stuttgart: Fischer 1988.

Zucker, L.G., Darby, M.R., Present at the biotechnology revolution: Transformation of technological identity for a large incumbent pharmaceutical firm, Research Policy, vol. 26, 1997, pp. 429-446.

List of contributors

Name	Address	Phone / Fax / e-mail
Bierly, Paul	James Madison University College of Business School of Professional Studies-Management MSC 0205 Harrisonburg, VA 22807	Phone: (540) 568-3236 Fax: (540) 568-2754 bierlype@jmu.edu
Brockhoff, Klaus	Institute for Research in Innovation Management University of Kiel Westring 425 D-24098 Kiel	Phone: +49 431 880 2165 Fax: +49 431 880 3349 brockhoff@bwl.uni-kiel.de
Chakrabarti, Alok K.	School of Management New Jersey Institute of Technology University Heights Newark, NJ 07102-1982	Phone: (973) 596-3256 Fax: (973) 596-3074 chakraba@megahertz.njit.edu
Damanpour, Fariborz	Department of Organization Management Faculty of Management Rutgers University 81 New Street Newark, NJ 07102-1820	Phone: (973) 353-5050 Fax: (973) 353-1664 fdamanpo@gsmack.rutgers.edu
Ernst, Holger	Institute for Research in Innovation Management University of Kiel Westring 425 D-24098 Kiel	Phone: +49 431 880 3614 Fax: +49 431 880 3349 ernst@bwl.uni-kiel.de

Farris, George F.	Management Research Center Faculty of Management Rutgers University 180 University Avenue Newark, NJ 07102-1895	Phone: (973) 353-5982 Fax: (973) 353-1664 gfarris@gsmack.rutgers.edu
Gemünden, Hans Georg	Inst. für Angewandte Betriebswirtschaftslehre und Unternehmensführung University of Karlsruhe P.O.Box 6980 D-76128 Karlsruhe	Phone: +49 721 608 3431 Fax: +49 721 608 6046 hans.gemuenden@wiwi.uni- karlsruhe.de
Gopalakrishnan, Shanthi	Dept. of Management, Marketing and ISS Fairleigh Dickinson University 1000 River Road Teaneck, NJ 07666	Phone: (201) 692-7231 Fax: (201) 692-7219 sgopalkr@alpha.fdu.edu
Harhoff, Dietmar	Inst. for Research on Innovation and Technology Management Ludwig-Maximilians- University of Munich, Ludwigstraße 28 D-80539 Munich	Phone: +49 89 2180 2239 Fax: +49 89 2180 6284 harhoff@bwl.uni-muenchen.de
Hauschildt, Jürgen	Institute for Research in Innovation Management University of Kiel Westring 425 D-24098 Kiel	Phone: +49 431 880 3999 Fax: +49 431 880 3213 hauschildt@bwl.uni-kiel.de
Kessler, Eric H.	Lubin School of Business Pace University 1 Pace Plaza New York, NY 10038	Phone: (212) 346-1846 Fax: (212) 346-1573 ekessler@pace.edu
Leker, Jens	Institute for Research in In- novation Management University of Kiel Westring 425 D-24098 Kiel	Phone: +49 431 880 3997 Fax: +49 431 880 3213

Markham, Stephen K.	Dept. of Business Management College of Management North Carolina State University Box 7229 Raleigh, NC 27695-7229	Phone: (919) 515-5592 Fax: (919) 515-5564 markham@comfs1.com.ncsu.edu